Springer Series in Advanced Microelectronics

Volume 42

For further volumes:
www.springer.com/series/4076

The Springer Series in Advanced Microelectronics provides systematic information on all the topics relevant for the design, processing, and manufacturing of microelectronic devices. The books, each prepared by leading researchers or engineers in their fields, cover the basic and advanced aspects of topics such as wafer processing, materials, device design, device technologies, circuit design, VLSI implementation, and subsystem technology. The series forms a bridge between physics and engineering and the volumes will appeal to practicing engineers as well as research scientists.

Geert Hellings · Kristin De Meyer

High Mobility and Quantum Well Transistors

Design and TCAD Simulation

Geert Hellings
CMOS Technology Department
IMEC
Leuven, Belgium

Kristin De Meyer
CMOS Technology Department
IMEC
Leuven, Belgium

ISSN 1437-0387 Springer Series in Advanced Microelectronics
ISBN 978-94-007-6339-5 ISBN 978-94-007-6340-1 (eBook)
DOI 10.1007/978-94-007-6340-1
Springer Dordrecht Heidelberg New York London

Library of Congress Control Number: 2013935397

Printed on acid-free paper

Springer is part of Springer Science+Business Media (www.springer.com)

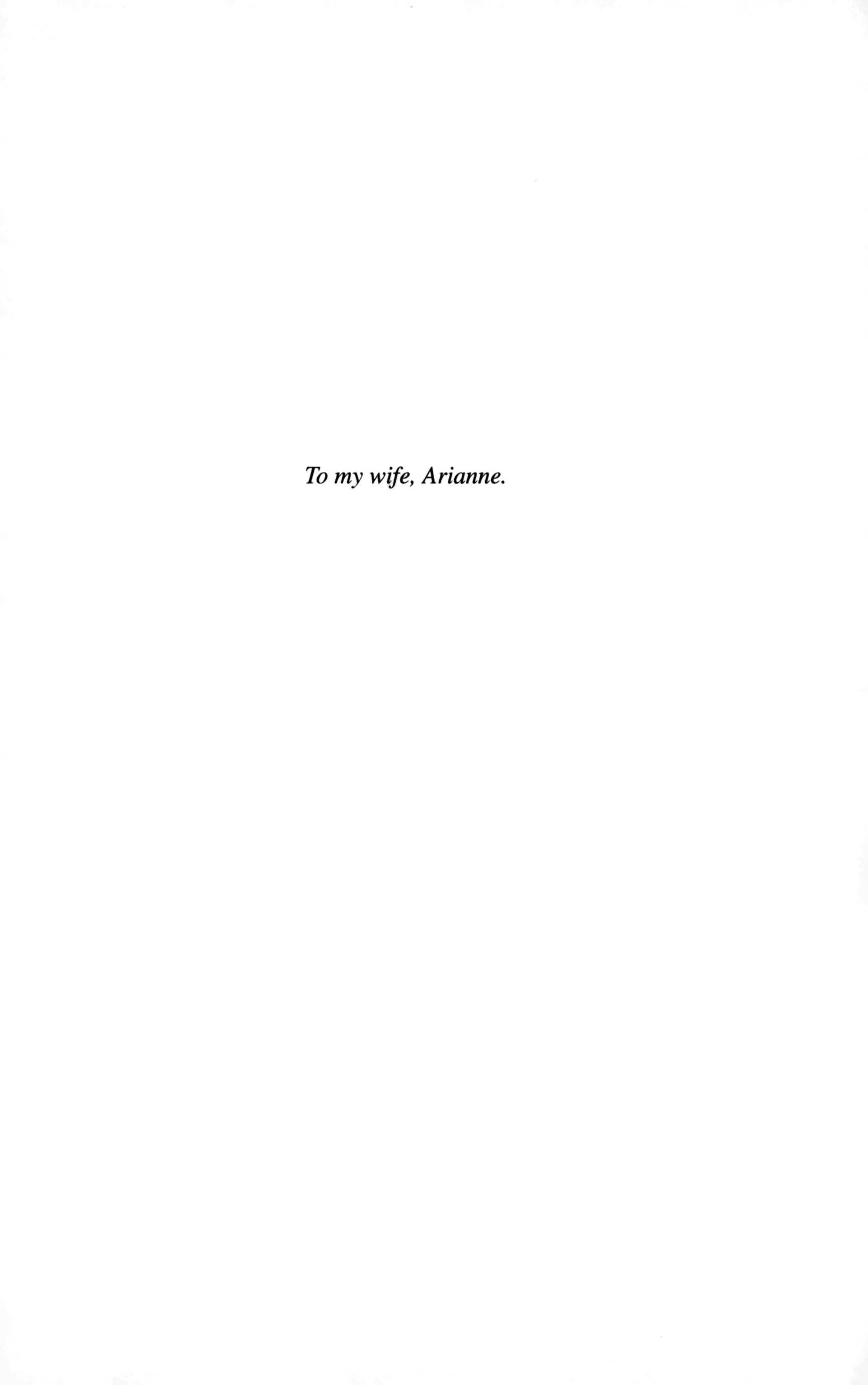

To my wife, Arianne.

Preface

The preface of a book a is generally considered to be a suitable place for philosophical thoughts '*There is no I in research.*' [88] '*Everything that has a beginning has an end.*' [31]. It is also a place to display creativity, sometimes in the original format (e.g. the '*Matrix of Gratitude*' [78]), sometimes in finding a dozen different ways to say *Thank you* [37, 145]. However, its ultimate function is to serve as a platform where the author can acknowledge invaluable contributions by others.

First of all, I would like to thank prof. dr. Kristin De Meyer, who gave me the opportunity to start a Ph.D. Her continuous support and encouragement have been extremely valuable. I appreciate her providing me the freedom to pursue many ideas. Allowing me a great deal of freedom, our many encounters made me think critically about my own research. Secondly, I am truly indebted to dr. Geert Eneman who served as the sounding board for many strange ideas. I also would like to acknowledge profs. Wim Dehaene, Marc Heyns, Marc Van Rossum, Guido Groeseneken, Siegfried Mantl and Paul Van Houtte for their careful review of the manuscript.

The seeds for this book were planted in fertile soil, commonly referred to as '*The EXPLORE Program*' at imec, Leuven, Belgium. To those early colleagues, I am grateful: Marc Meuris, Marc Heyns, Brice De Jaeger, David Brunco, Jerome Mitard, Eddy Simoen, Ali Pourghaderi, Gillis Winderickx and Koen Martens. In those 'early days', the combined decades of experience surrounding my desk were a true inspiration: An De Keersgieter, Isabelle Ferain, Stefan Kubicek, Nadine Collaert, Rita Rooyackers.

A big thank you also goes to those who later crossed my path: Niamh Waldron, AliReeza Alian, Liesbeth Witters, Roger Loo, Andriy Hikavyy, Thomas Hoffman, Hugo Bender, Trudo Clarysse, Raymond Krom, Dennis Lin, Taiji Noda, Aftab Nazir, Andreas Schulze, Jacopo Franco, Luigi Pantisano, Wei-E Wang, Romain Ritzenthaler, Shinpei Yamaguchi ang Gerd Zschaetzsch. Thank you all for answering numerous questions and for all the joint publications.

The contributions of many experts are much appreciated: prof. Asen Asenov and Brahim Benbakhti (University of Glasgow, UK), Karol Kalna (Swansea University, UK), Athanasios Dimoulas (NCSR Demokritos, Greece), Clemens Wuendisch and

Matthias Posselt (HZ Dresden Rossendorf, Germany), Dirch Hjorth Petersen (DTU Nanotech, Kgs. Lyngby, Denmark).

The authors also acknowledges the European Commission for financial support in the DualLogic procect no. 214579 as well as imec and its core partners within the Industrial Affiliation Program on Logic/DRAM and the Institute for the promotion of innovation in Flanders (IWT-Vlaanderen).

To my friends, I would like to say thank you as well, for the many nice moments including MovieKebaps (is that a real word?), poker nights, barbecues and the occasional cocktail party. I will not attempt to list all of you here, since such efforts are futile by definition. I would like to thank my family and parents in particular, for giving me the chance to start studying and to support me in achieving my goals, scientific and other. Finally, a big thank you also goes to my wife, Arianne, for her continuous unconditional love and support.

Leuven, Belgium Geert Hellings

Abstract

For many decades, the semiconductor industry has miniaturized transistors, delivering increased computing power to consumers at decreased cost. However, mere transistor downsizing does no longer provide the same improvements. One interesting option to further improve transistor characteristics is to use high mobility materials such as germanium and III-V materials. However, transistors have to be redesigned in order to fully benefit from these alternative materials. This book shows that Quantum Well based transistors can be quite suited for this, since they confine the charge carriers to the high-mobility material using a heterostructure. One particular structure, the SiGe Implant-Free Quantum Well pFET was designed and fabricated using industry-scaled infrastructure. Electrical testing showed remarkable short-channel performance and prototypes were found to be competitive with a state-of-the-art planar strained-silicon technology. High mobility channel, providing high drive current, and heterostructure confinement, providing good short-channel control make an interesting combination for future technology nodes.

Starting with an investigation of a bulk germanium FET technology, the fabrication of shallow junctions in germanium substrates was investigated. Both boron and gallium were found to be suitable p-type dopants, delivering high electrical activation up to 4×10^{20} cm^{-3} while dopant diffusion remains negligible (under certain conditions). Considering n-type dopants, arsenic was studied, focussing on millisecond laser annealing in an attempt to reduce the concentration-enhanced diffusion. While the active dopant concentration was rather high, significant diffusion was still observed.

Following this experimental work, a Monte Carlo simulator was calibrated to enable TCAD simulations of ion implants into Ge substrates. Simulated as-implanted profiles for B, P, Ga and As showed good agreement with experimental data. Using this calibrated MC simulator, the ion implant steps for a scaled 70-nm Ge pMOSFET technology were designed. Fabricated transistors were found to outperform the ITRS requirements for the corresponding technology node. A commercial TCAD simulator was also extended to allow electrical simulations of Ge pMOSFETs. Specifically, models for carrier mobility and generation/recombination processes were calibrated using experimental data. Electrical simulations of Ge pMOSFETs were found to be in good agreement with electrical measurements. Typical performance metrics (I_{ON}, I_{OFF}, *DIBL* etc.) were within 5–10 % of experimental values. Complementing experimental work, this TCAD combination allows optimizing and predicting the performance of new, scaled germanium-based devices.

However, it seems unlikely that a planar bulk germanium technology would be well suited for future technology nodes, because of drain-to-bulk junction leakage. Addressing this issue, another strategy was followed to integrate high-mobility channel materials such as germanium into future technology nodes. A class of transistors was introduced, which only uses the high-mobility material in the transistor channel. They are designed in such a way that the charge carriers are confined to a Quantum Well by means of heterostructure confinement.

In the Si/SiGe material system, the SiGe Implant-Free Quantum Well (IFQW) transistor was developed and TCAD simulations predicted excellent short channel control down to 16 nm gate lengths, markedly better than for equivalent bulk silicon pFETs. In InGaAs-based IFQW nFET was also designed, showing good short channel control at a gate length of 10 nm.

Finally, the SiGe Implant-Free Quantum Well transistors were fabricated using industry-scaled infrastructure. First-generation SiGe IFQW pFETs with raised source and drain were electrically analyzed. Devices with gate lengths down to 30 nm showed excellent short channel control with *DIBL* and *SS* values of 126 mV/V and 80 mV/dec respectively. Compared to Si control pFETs, a 50 % higher drive current was obtained. Integrating embedded SiGe source drain stressors into this IFQW pFET architecture, second-generation IFQW pFETs were fabricated. These prototypes were found to be competitive with a 32-nm node state-of-the-art strained-silicon technology, combining a high saturation drive current of 1 μA/μm, maintaining the improved short channel control. Considering that there is still significant room for further improvement of this IFQW pFET, this comparison suggests that it should be considered a viable technology option for upcoming technology nodes.

Contents

List of Abbreviations

a/c	Interface between amorphous and crystalline material
α-Ge	Amorphized Germanium
c-Ge	Crystalline Germanium
APT	Anti Punch Through Implant (mid-range well implant)
CMOS	Complementary Metal-Oxide-Semiconductor
δ-doping	Delta Doping layer
FLA	Flash Anneal
4PP	Four-Point Probe
HAADF-STEM	High Angle Annular Dark Field Scanning Transmission Electron Microscopy
HDD	Highly Doped Drain
I/I	Ion Implant
ITRS	International Technology Roadmap for Semiconductors
LDD	Lowly Doped Drain (extension regions)
LSA	Laser Anneal
μ4PP	Micro Four-Point Probe
MOSFET	Metal-Oxide-Semiconductor Field Effect Transistor
PAI	Pre-Amorphization Implant
RTA	Rapid Thermal Anneal
SE	Spectroscopic Ellipsometry
SIMS	Secondary-Ion Mass Spectroscopy
SPER	Solid Phase Epitaxial Regrowth
SRP	Spreading Resistance Probe
TCAD	Technology Computer Aided Design
TEM	Transmission Electron Microscopy
USJ	Ultra Shallow Junction
VLSI	Very Large Scale Integration
VPS	Variable Probe Spacing
VTA	Threshold Voltage Adjust Implant (shallowest well implant)

List of Symbols

β_p	–	Kurtosis (4th moment of the Pearson distribution curve)
C_{BC}	F/cm^2	Bulk-to-Channel Capacitance
CET	nm	Capacitance equivalent thickness
C_{GC}	F/cm^2	Gate-to-Channel Capacitance
C_{OX}	F/cm^2	Oxide Capacitance
C_{INV}	F/cm^2	Transistor Inversion Capacitance
ΔE_C	eV	Conduction Band Energy Offset (between 2 materials)
ΔE_V	eV	Valence Band Energy Offset (between 2 materials)
$\Delta\Psi_{OX}$	eV	Potential Drop over the Dielectric
$\Delta\Psi_S$	eV	Potential Drop over the semiconductor
$DIBL$	mV/V	Drain Induced Barrier Lowering
D_{IT}	cm^{-2} eV^{-1}	Density of Interface Traps
EOT	nm	Equivalent oxide thickness
ϵ_S	F/m	Substrate relative permittivity
ϵ_{XX}	%	Longitudinal Relative Strain
ϵ_{YY}	%	Transversal Relative Strain
ϵ_{ZZ}	%	Vertical Relative Strain
E_C	eV	Conduction Band Energy
E_F	eV	Fermi Energy Level
E_G	eV	Semiconductor Band Gap
E_V	eV	Valence Band Energy
γ_p	–	Skewness (3rd moment of the Pearson distribution curve)
$g_{m,max}$	A/V	Transistor Transconductance
h_{sp}	nm	Spacer Height
I_B	A	Transistor Bulk Current
I_D	A	Transistor Drain Current
I_{DS}	A	Transistor Drain-to-Source Current
I_G	A	Transistor Gate Current
I_{ON}	µA/µm	Transistor ON-state Current
I_{EFF}	µA/µm	Transistor Effective ON-state Current

I_{OFF}	A/µm	Transistor OFF-state Current
I_S	A	Transistor Source Current
k_b	$m^2\,kg\,s^{-2}\,K^{-1}$	Boltzmann Constant
L_G	nm	Transistor Gate Length
$Ł_\delta$	nm	Length over which the δ-doping is interrupted
μ	cm^2/Vs	Mobility
μ_H	cm^2/Vs	Hall Mobility
μ_0	cm^2/Vs	Intrinsic Channel Mobility
N_{ACT}	cm^{-3}	Active Carrier Concentration
N_{INV}	cm^{-2}	Number of Mobile (inversion) Charge Carriers
N_{WELL}	cm^{-3}	Well Doping Concentration
Φ_B	eV	Schottky Barrier Height
Φ_{MS}	eV	Metal-to-Semiconductor Workfunction Difference
q	C	Elementary Charge
Q_{INV}	C/cm^2	Inversion Charge
Q_{SS}	C/cm^2	Charge density related to surface states
R_{EXT}	Ω	Series Resistance
$^{-1}$	nm	Range (1st moment of the Pearson distribution curve)
R_{sh}	$\Omega/sq.$	Sheet Resistance
σ_p	nm	Straggle (2nd moment of the Pearson distribution curve)
σ_{XX}	Pa	Longitudinal Stress
σ_{YY}	Pa	Transversal Stress
σ_{ZZ}	Pa	Vertical Stress
SS	mV/dec	Subthreshold Swing
T_{melt}	K	Melting Temperature
T_{peak}	K	Peak Wafer Temperature (during LSA)
V_B	V	Transistor Bulk Voltage
V_D	V	Transistor Drain Voltage
V_{DD}	V	(Digital Logic) Supply Voltage
V_{DS}	V	Transistor Drain-to-Source Voltage
V_{FB}	V	MOSCAP/MOSFET Flatband Voltage
V_G	V	Transistor Gate Voltage
V_{GS}	V	Transistor Gate-to-Source Voltage
V_S	V	Transistor Source Voltage
v_{sat}	cm/s	Electron/Hole Saturation Velocity
V_T	V	Transistor Threshold Voltage
V_{FB}	V	MOSCAP/MOSFET Threshold Voltage (onset of inversion)
$V_{T,sat}$	V	Transistor Saturation Threshold Voltage
X_J	m	Junction depth
Z_D	m	Depletion depth

Chapter 1
Introduction

1.1 Transistor Scaling

This section briefly discusses the historical context and major technological developments in the microelectronics industry.

1.1.1 The Early Days (1925–1960)

In 1927, the Polish physicist J.E. Lilienfeld was granted what is believed to be the first patent for the Field-Effect Transistor (FET) principle in Canada [87]. However, it took another two decades before Shockley, Bardeen and Brattain delivered experimental results. They observed that when two gold point contacts were applied to a crystal of germanium, a signal was produced of which the output power exceeded the input power. Another decade later in 1958, Kilby produced the first integrated circuit (IC), commonly known as a *chip*, also using germanium. The first Metal-Oxide-Semiconductor FET (MOSFET) was made in 1960 by Kahng and Atalla. Unlike their predecessors, they used silicon crystals. The reason for this was that Atalla had shown that a high-quality Si/SiO_2 interface could be obtained using his oxidation process. This high-quality Si/SiO_2 interface was extremely important, since it constitutes the heart of the MOSFET. In contrast, the oxide of germanium is soluble in water and its interface with germanium often contains plenty of defects. By the end of the 1960's, silicon had become the dominant semiconductor.

1.1.2 The Happy Days of Scaling (1960–2000)

It was Gordon Moore [103] in 1965, who observed that the number of transistors on an integrated circuit (for equal component cost) doubles every 2 years. This empirical law has indeed been the driving force behind the semiconductor industry for

G. Hellings, K. De Meyer, *High Mobility and Quantum Well Transistors*,
Springer Series in Advanced Microelectronics 42, DOI 10.1007/978-94-007-6340-1_1,

many decades. Throughout this period, the architecture and working principle of the MOSFET have essentially not been changed. The continuous scaling of its physical dimensions has delivered improved performance and lower cost at every single technology generation.

Until 2000, this evolution has driven the growing computing power of the PC and ever increasing functionalities (more data storage, performing games, internet access, etc.). This era is referred to as *The Happy Days of Scaling*. Scaling laws have been discussed extensively in literature. As such, they will not be discussed again in this introduction. However, in recent years the downsizing of the components no longer guarantees the combined bonuses of higher performance and lower cost. The happy scaling days are over! [30].

1.1.3 Materials-Based Scaling (2000–2010)

In the year 2000, silicon MOSFETs had been reduced in size by a several orders of magnitude. The transistor gate length L_G had been reduced to about 100 nm, a distance corresponding to approximately 290 silicon atoms. Beyond this point, it became increasingly difficult to achieve the desired transistor characteristics. Short channel effects make it increasingly difficult to control the OFF-state current and to keep the MOSFET's power consumption at an acceptable level. Thin gate dielectrics lead to higher gate leakage and can severely impact the long-term reliability of a technology. Finally, higher doping levels and increased electric fields result in a reduced channel mobility for successive technology generations, harming the drive current. In short, making transistors smaller delivers only a limited improvement. Instead, OFF-state leakage current increases power consumption while the boost in ON-current becomes smaller.

This observation has proven to be a strong stimulus to incorporate many new materials into the VLSI circuitry. Where the typical 1990 MOSFET mainly consisted of 3 materials (silicon, silicondioxide, and aluminum), a 2010 MOSFET contains many more. silicongermanium has been introduced in the source-drain regions to strain the channel, hereby increasing ON-state drive current. Dielectrics with a higher permittivity than SiO_2 such as SiON, HfO_2, etc. have allowed a further increase of the gate-to-channel capacitance, without making the dielectric thinner. Many kinds of metal gates have assisted in alleviating the gate depletion problem and in reducing the gate electrode's resistance. Partial silicidation of the source/drain contacts using cobalt, nickel, platinum, etc. has reduced MOSFET series resistance, while copper and tungsten have replaced aluminum in the interconnect layers and via plugs. While the list is probably even longer, it should be obvious that MOSFET technology has put a big part of Mendeliev's periodic table to work. By doing this, Moore's law has been pushed further, with present-day MOSFETs having gate lengths of only 20 nm. However, while many of these innovations to the MOSFET solve one problem, they often introduce other issues.

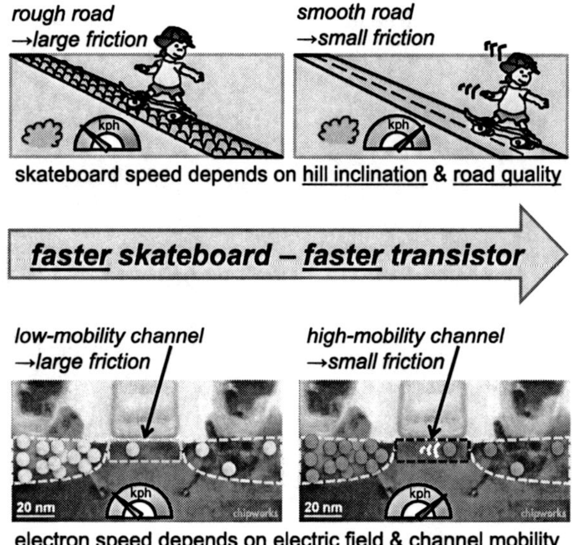

Fig. 1.1 Gliding downhill on a skateboard, the velocity is limited by the quality of the road. Similarly, the velocity at which an electron transverses the transistor channel under the influence of an electric field, is limited by the electron mobility of the channel material (Transistor magnification: 500000×)

1.2 What's Next? (2010–...)

This section discusses two options to extend transistor scaling to future technology generations, that will be the topic of this book.

1.2.1 High-Mobility Channel Materials

The first option is sometimes referred to as *Performance Scaling*: transistor characteristics are improved by improving the intrinsic properties of the transistor channel. This allows to achieve a boost in performance, without requiring a size reduction of the MOSFET. As mentioned before, the main reason to choose silicon above other semiconductors is the intrinsic stability of the Si/SiO_2 interface. However, this argument is less relevant when the gate dielectric of choice is something else than SiO_2. As a result, different semiconductors are now also being considered. Semiconductors with a higher mobility than silicon offer the possibility to obtain higher drive currents than can be achieved in a technology using a silicon channel. Since germanium and many III-V semiconductors (compounds consisting of elements from group III and group V in the periodic table) offer a higher mobility than silicon, these are excellent candidates to replace silicon in future technologies.

This idea is shown schematically in Fig. 1.1 and explained using an analogy with skateboarding downhill. For this system, the velocity is influenced by the inclination and the road quality (these are resp. driving force and friction force). Obviously, a small inclination and a bumpy road will result in slow skateboarding. In a transistor channel, the situation is not much different: electrons are propelled across a transistor's channel by the electric field (driving force, equivalent to hill inclination), and slowed down by the crystal lattice in which they travel (friction force, equivalent to road quality). High-mobility materials such as germanium and many III-V materials allow faster movement of electrons, resulting in faster circuits. Just as faster skate-boarding gets you to the bottom of the hill sooner, faster transistors get a computer to the end of that long calculation faster.

1.2.2 Scalable Transistor Designs

The second option is to change the *structure* of the transistor, so that short channel effects are avoided. In this context, architectures such as the Multiple-Gate FETs (MuGFETs) or Silicon-On-Insulator FETs (SOI-FETs) are commonly considered. Both improve the transistor's characteristics by reducing the short channel effects that have become problematic in planar silicon technology. In the MuGFETs, the charge carriers are confined to a thin free standing silicon rib (often referred to as *fin*), which is surrounded by the gate electrode on both sides. In SOI-FETs a buried SiO_2-layer is placed below a thin silicon channel connecting source and drain. As such, both transistor architectures interrupt common OFF-state leakage paths found in planar bulk silicon MOSFETs. Needless to say, both options constitute a drastic change in the transistor morphology.

1.3 Goals of the Book

The general objective of this book is to evaluate the benefits and limitations of a technology that uses high-mobility channel materials, and to design transistors that fully benefit from higher mobility.

For this technology evaluation, experimental work will be combined with finite-element based simulation programs (Technology Computer Aided Design or TCAD). The latter have proven to be extremely valuable in the development of silicon MOSFETs during the last decades. Given the fact that germanium technology is less mature than the industry-standard silicon, calibrated and trustworthy TCAD simulations are considered indispensable to fully assess the benefits and limitations of germanium technology. Therefore, the first goal is to implement the necessary physical models to allow for trustworthy TCAD simulations of scaled germanium-based MOSFETs. Such a simulator would be a valuable tool in designing transistor structures that can deliver improved performance for the upcoming technology nodes.

A second goal of this book is to design and fabricate scalable transistors using these high-mobility materials, thus combining both options discussed in the previous section. Obviously, any new transistor design should be thoroughly evaluated and benchmarked to existing state-of-the-art silicon technology.

1.4 Organization of the Book

In Chap. 2, the fabrication of shallow junctions in germanium substrates is discussed. Considering p-type junctions, boron and gallium are studied, focussing on electrical activation and diffusion behavior. Considering n-type junctions, arsenic is studied, focussing on millisecond laser annealing in an attempt to reduce the concentration-enhanced diffusion and resulting arsenic-deactivation commonly observed using classical activation anneals. Finally, this chapter intends to explore opportunities for junctions in high mobility materials by benchmarking against existing literature data and against the ITRS requirements for the upcoming technology nodes.

In Chap. 3, a Monte Carlo simulator is calibrated to enable TCAD simulations of ion implants into germanium substrates. Using this calibrated simulator, the ion implant steps required for a scaled $L_G = 70$ nm germanium pMOSFET technology are designed. These devices are fabricated and electrically analyzed and benchmarked against silicon pFETs from the corresponding technology node.

In Chap. 4, a TCAD device simulator is extended to allow electrical simulations of germanium pMOSFETs. Using this tool and the calibrated ion implant simulator, electrical simulations of germanium pMOSFETs with L_G ranging from 70 nm to 1 μm are calibrated against experimental results. Building on these TCAD capabilities, a methodology is presented allowing to study and predict the effect of interface traps in germanium technology on transistor performance. Finally, the impact of interface traps on MOSFET drive current is investigated.

In Chap. 5, scaling issues and short channel effects in the bulk MOSFET are analyzed using a TCAD simulator. Particularly, the impact of drain-to-bulk junction leakage in a bulk $L_G = 65$ nm germanium MOSFET technology is investigated. Secondly, a class of transistors is introduced, where charge carriers are confined to a Quantum Well (QW) by means of heterostructure confinement. The Implant-Free Quantum Well FET is presented and its scaling performance analyzed for gate lengths down to 16 nm using the silicon/silicongermanium material system as an example. Zooming in on the critical interface between the QW channel and the underlying substrate, the role of band offsets is discussed in detail. An InGaAs/GaAs IFQW nFET is also introduced, starting from a classical High Electron Mobility Transistor (HEMT). Finally, analytical expressions are derived for the depth of the depletion layer, the threshold voltage and the body factor in Quantum Well MOSFETs.

In Chap. 6, following up on the TCAD simulations in the preceding chapter, the process development to fabricate silicongermanium-based IFQW pFETs are treated. *First-generation* silicongermanium IFQW pFETs are fabricated and electrically analyzed, focussing on their enhanced scalability. Integrating embedded $Si_{0.75}Ge_{0.25}$ source/drain stressors into this IFQW pFET architecture, *Second-generation* silicongermanium IFQW pFETs are also fabricated and compared to $L_G = 32$ nm state-of-the-art planar strained-silicon technology. The performance at lower operating voltage is shown and compared with strained-SOI nFETs. Finally, the matching performance and V_T-tuning capabilities of IFQW pFETs is explored.

Chapter 2
Source/Drain Junctions in Germanium: Experimental Investigation

In this chapter, the fabrication of shallow junctions in germanium is investigated experimentally, targeting application in a scaled germanium MOSFET technology.

2.1 Introduction

Improvements in the microelectronic industry over the past decades have relied heavily on a continuous effort to overcome the difficulties of device shrinking [75]. Germanium is considered as an attractive high mobility substrate for high performance CMOS applications. One critical issue towards the establishment of a scaled Ge-based technology is the ability to fabricate shallow, low-resistive junctions in this material. Consequently, a systematic investigation of the diffusion and activation behavior of ion-implanted (I/I) dopants in Ge substrates during drive-in anneal is required.

Regarding p-type junctions, existing studies have focussed mainly on boron, also the dominant p-type dopant in Si technology. An interesting alternative for p-type doping of germanium is gallium: its higher atomic mass reduces both the straggle and ion-channeling during I/I, leading to a more abrupt as-implanted profile (compared to the lighter B). For this reason, the implantation of Ga and its behavior during subsequent annealing will be investigated in this chapter. In a second study, ion-implanted boron in Ge will be studied in sub-30 nm junctions.

Regarding n-type junctions, phosphorus and arsenic both suffer from significant concentration-enhanced diffusion in germanium through the formation of arsenic or phosphorus-Vacancy complexes, at concentrations in excess of $2\text{–}5 \times 10^{19}\ \text{cm}^{-3}$. As such, the fabrication of ultra-shallow n-type junctions in Ge requires limiting the activation anneal's thermal budget while at the same time, a high temperature is required to achieve a high electrically active concentration. Attempting to combine these two competing requirements, the feasibility of millisecond laser annealing (LSA) to fabricate ultra-shallow low-resistive n-type junctions is investigated, using As.

G. Hellings, K. De Meyer, *High Mobility and Quantum Well Transistors*,
Springer Series in Advanced Microelectronics 42, DOI 10.1007/978-94-007-6340-1_2,

Finally, this chapter intends to explore opportunities for junctions in high mobility materials by benchmarking against existing literature data and against the ITRS requirements for the upcoming technology nodes.

2.2 p-Type Junctions

2.2.1 Furnace Annealed Gallium Junctions

The goal of this section is to investigate the behavior of ion-implanted gallium and its subsequent annealing at different temperatures in preamorphized (α-Ge) and crystalline germanium (c-Ge). To this end, the as-implanted profiles, electrical activation, diffusion and recrystallization process will be discussed.

2.2.1.1 Experimental Details

Czochralski-grown 100 mm diameter, 350 μm-thick, (100)-oriented, n-type bulk Ge wafers were obtained from Umicore. On one set of wafers, a Preamorphization Implant (PAI) was performed using Ge at an energy of 200 keV and with a dose of 10^{15} cm^{-2}. This PAI results in a 190 nm-thick amorphized layer on these wafers. Subsequently, Ga was implanted at different energies (40 and 80 keV) with a dose of 3×10^{15} cm^{-2} with a 7° tilt with respect to the wafer normal. On another set of wafers, Ga was directly implanted without a preceding PAI-step. Immediately after the I/I, all wafers were capped with a protecting 20 nm SiO_2 layer. The annealing was performed in N_2 atmosphere in a *Heatpulse 610* rapid thermal annealing (RTA) system at temperatures between 300 and 700 °C for 60 s. The chemical dopant concentrations were studied by Secondary-Ion Mass Spectroscopy (SIMS). The electrical activation of Ga was analyzed by sheet resistance measurements with a conventional four-point probe (4PP) system, the variable probe spacing technique (VPS, [25]) and a micro-four-point probe system (μ4PP, [112]). The implant-induced damage and residual disorder after annealing were studied with transmission electron microscopy (TEM).

2.2.1.2 Physical Characterization

Figure 2.1 shows the as-implanted Ga profiles with energies of 40 and 80 keV and a dose of 3×10^{15} cm^{-2} in both α-Ge and c-Ge. In the c-Ge, a channeling tail is clearly distinguishable at concentrations below 10^{18} cm^{-3}. The as-implanted profiles (in α-Ge) were fitted with Pearson distribution curves [5] of which the resulting moments are also given.

The implantation of 40 keV Ga with a dose of 3×10^{15} cm^{-2} amorphized the c-Ge to a depth of 58 nm, as shown by the TEM micrograph in Fig. 2.2(a).

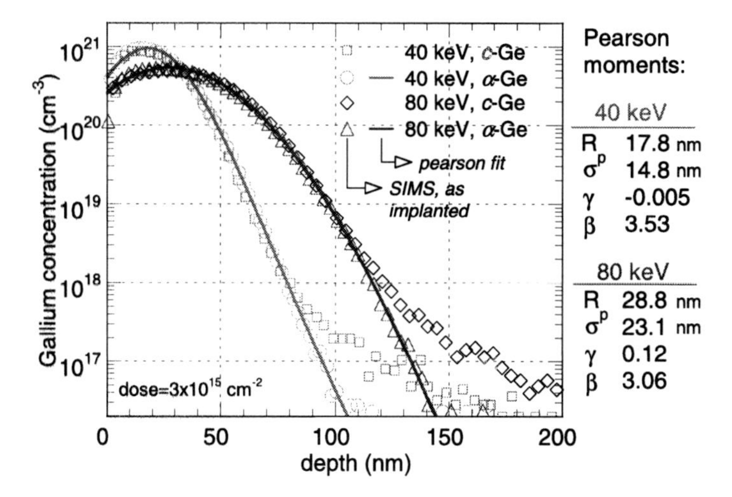

Fig. 2.1 Chemical concentration profiles as a function of depth for Ga, as implanted in the α-Ge and c-Ge samples (40 and 80 keV 3×10^{15} cm^{-2}). The profiles in α-Ge were also fitted with Pearson distribution curves

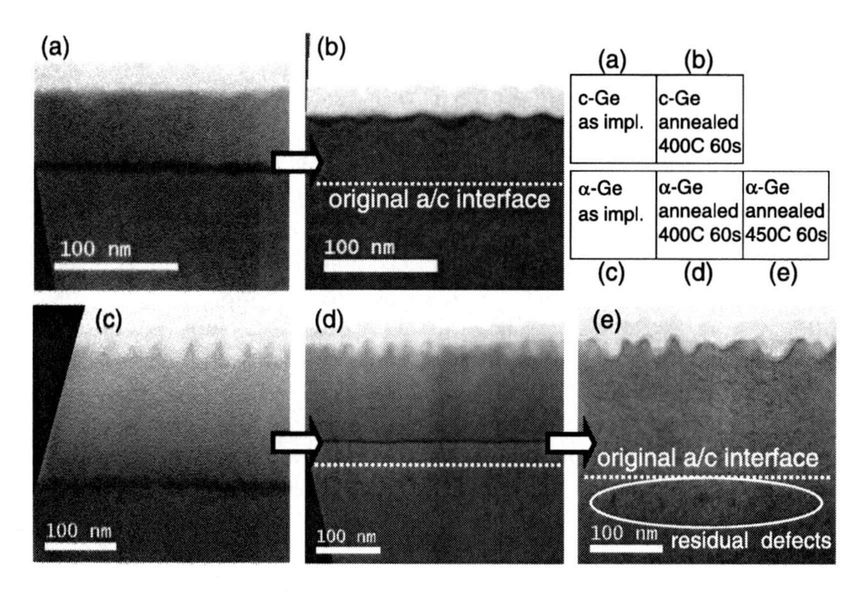

Fig. 2.2 Cross-sectional TEM of 40 keV 3×10^{15} cm^{-2} Ga implanted in c-Ge (*top*) and α-Ge (*bottom*)

The 80 keV implant is expected to yield an amorphous layer of 92 nm, based on Monte Carlo simulations—see Chap. 3). Full recrystallization occurs after the 60 s anneal at 400 °C (Fig. 2.2(b)). The Ge surface appears to be quite rough, even before the recrystallization. This can be attributed to vacancy migration [76] because the I/I was performed prior to depositing the protecting SiO_2 cap layer. While this

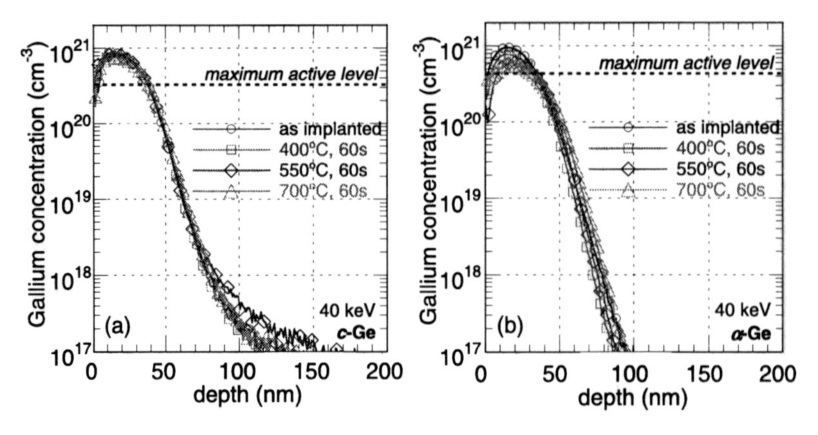

Fig. 2.3 Chemical concentration profiles as a function of depth for Ga (40 keV 3×10^{15} cm^{-2}), as-implanted and annealed at 400, 550 and 700 °C for 60 s in c-Ge and α-Ge

surface roughening for high-dose, heavy-ion I/I at room temperature is undesirable for device fabrication, it can be avoided by cooling the substrate during the I/I-step (hence suppressing the vacancy diffusion). In the α-Ge samples, the roughening is even more severe due to the 200 keV Ge PAI, as shown in Fig. 2.2(c). Here, only 40 nm of Ge was regrown during the 400 °C anneal (Fig. 2.2(d)), indicating a slower regrowth than in the c-Ge sample and as reported by Csepregi et al. in [29]. Full recrystallization however was achieved with the 450 °C anneal, while residual crystal damage can still be observed below the original a/c interface, agreeing with the observations by Hickey et al. in [65] (Fig. 2.2(e)).

To investigate the diffusion behavior, the Ga profile was analyzed in SiO_2-capped samples annealed at 400, 550 and 700 °C for 60 s using SIMS. The corresponding profiles are presented in Fig. 2.3(a) and (b) for c-Ge and α-Ge substrates respectively. As shown in Fig. 2.3(a), no diffusion could be observed in the c-Ge sample for temperatures up to 700 °C, confirming the reduced diffusivity of group III elements in Ge reported in [120]. Within the accuracy of the SIMS analysis, the total implanted dose was fully retained in the sample during the anneal. Also in the α-Ge samples, no diffusion was observed up to 700 °C. However, the peak Ga concentration dropped steadily from 9×10^{20} to 5.5×10^{20} cm^{-3} with increasing temperature. In the 550 °C sample, only 70 % of the total implanted dose was retained (no additional dose loss was observed by further increasing the RTA temperature up to 700 °C).

To find an explanation for this, the recrystallization was analyzed in more detail. The thicker amorphous layer (190 nm) and the observed slower crystal regrowth imply that Ga remains longer in the amorphous phase for the α-Ge samples. This also implies that Ga is still in the amorphous phase at higher temperatures during the ramp-up (rate 20 °C/s) of the high-temperature RTA anneals. Illustrating this, the recrystallization process was calculated in more detail, using the temperature-dependent regrowth model from [29] (accounting for the slower regrowth observed

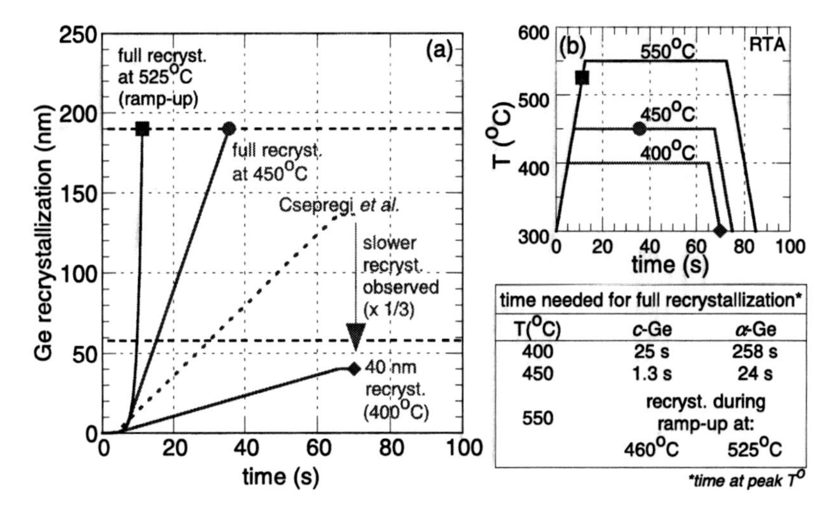

time needed for full recrystallization*		
T(°C)	*c*-Ge	*α*-Ge
400	25 s	258 s
450	1.3 s	24 s
550	recryst. during ramp-up at: 460°C	recryst. during ramp-up at: 525°C

*time at peak T°

Fig. 2.4 (**a**) Calculated recrystallization thickness as a function of time for the α-Ge samples for (**b**) three different anneal temperatures. The regrowth model from Csepregi et al. [29] was modified to account for the slower regrowth observed in our α-Ge samples (see text). Full recrystallization is expected to occur after 24 s (450 °C anneal) or during the ramp-up phase of the 550 °C anneal. The symbols link the end of each regrowth curve in (**a**) to a time-T° point in (**b**). Finally, *a table* shows the estimated annealing time required for full recrystallization

during the TEM analysis in the α-Ge samples). The results of these calculations are shown in Fig. 2.4:

- The 400 °C RTA anneal resulted in a 40 nm recrystallization based on the TEM analysis of the α-Ge samples. At this rate, an annealing time of 258 s would be required for full recrystallization of the 190 nm thick amorphous Ge layer. For the c-Ge sample, the 400 °C RTA anneal resulted in full recrystallization of the 58 nm (see Fig. 2.2(b)). Assuming the regrowth velocity from [29], one can estimate that the amorphous layer is fully recrystallized after only 25 s at 400 °C.
- The 450 °C RTA anneal resulted in a full recrystallization after 24 s for the α-Ge sample, in contrast to only 1.3 s for the c-Ge sample, where the amorphous layer almost fully recrystallizes during the ramp-up towards 450 °C.
- The 550 °C RTA anneal achieves full recrystallization of the amorphous layer during the ramp-up phase. This occurs at a temperature of 525 °C for the α-Ge sample and at 460 °C for the c-Ge sample.

Note that these estimates assume a 3× slower regrowth velocity for the α-Ge samples, as observed in our samples. Without this assumption, the difference between the c-Ge and α-Ge recrystallization times would be smaller (although still substantial).

With these considerations in mind, enhanced Ga diffusion in the amorphous Ge phase explains our observations: during the 400 °C anneal, Ga is present in the amorphous Ge phase at the same temperature (i.e. 400 °C) in the α-Ge and c-Ge samples. This corresponds with the small dose loss observed for the α-Ge sample

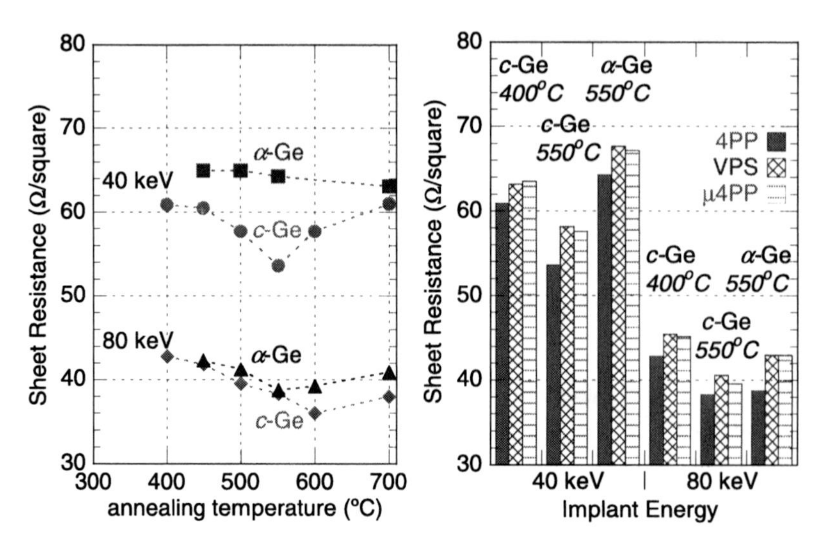

Fig. 2.5 (**a**) Sheet resistance of Ga junctions formed by I/I (40 and 80 keV) in α-Ge and c-Ge, as a function of annealing temperature measured with the conventional 4PP. (**b**) Sheet resistance for selected samples measured with the conventional 4PP, the VPS technique and the μ4PP system

(Fig. 2.3(b)), which can be attributed to the longer time spent at this temperature. During the higher temperature anneals, Ga remains in the amorphous Ge phase until the wafer reaches a temperature of 525 °C in α-Ge substrates, in contrast to only 460 °C for c-Ge. The fact that no additional dose loss is observed for a 700 °C anneal, in comparison to the 550 °C one, also indicates that the observed dose loss is linked to the identical feature of these two anneals, i.e. the ramp-up phase of the RTA-anneal. However, to fully confirm this model, additional dedicated experiments would be needed to rule out any effect related to the surface morphology (e.g. roughness).

2.2.1.3 Electrical Characterization

The electrical activation of Ga-doped Ge was studied by sheet resistance measurements (R_{sh}) using three techniques (4PP, VPS and μ4PP). While the relative simplicity of the conventional 4PP tool is an advantage, the probe penetration in Ge can be quite high (up to about 200 nm, depending on probe pressure [26]). Since the Ga junctions under investigation are considerably shallower than 200 nm, the concern is that the probe-needles would pierce the junction, yielding erroneous results. This was indeed found to be the case since no reliable, reproducible R_{sh} could be obtained using the conventional 4PP technique on our samples. Attempting to circumvent this, a second set of 4PP measurements was carried out while leaving the protecting SiO_2 cap layer on the sample. Apparently, this extra layer reduced the 4PP probe penetration enough to allow reproducible R_{sh} measurements: Fig. 2.5(a) shows the annealing temperature dependence of R_{sh}, for the Ga junctions fabricated in α-Ge and c-Ge. The sheet resistance remains rather constant for both implant energies as soon as the samples are fully recrystallized: the R_{sh} variation over the entire temperature range is only 10 %, with an optimum around 550 °C. This implies that a high Ga activation is already obtained at rather low temperatures (400 °C). Finally, the lowest R_{sh} value is obtained at a temperature of 500–600 °C.

In order to confirm the validity of our R_{sh} measurements with the 4PP-tool (leaving the SiO_2 layer present), VPS and μ4PP measurements were as well performed on selected samples (without the SiO_2 cap). The R_{sh} values are found to be consistent across these three techniques and results are plotted in Fig. 2.5(b): the difference between the VPS and μ4PP values is generally smaller than a few percent, while the 4PP measurements yield the same trend but give slightly lower values (5 %).

From the SIMS analysis of the annealed samples, the VPS-based R_{sh}, and a concentration-dependent mobility model for Ga-doped germanium [50], an active concentration level can be calculated. This calculation yields a maximum active concentration level of 4.4×10^{20} and 3.3×10^{20} cm^{-3} for Ga implanted into c-Ge and α-Ge respectively (the methodology of this calculation is explained in [123]). The accuracy of these levels is mainly determined by the uncertainty on the mobility model used, which dates back to 1962. A more recent mobility model is available for B-doped Ge [98], yielding slightly lower maximum active concentration levels (2.6×10^{20} cm^{-3}). However, Hall-mobility measurements carried out on our own samples tend to agree with the original model [63]. Finally, the observed active Ga concentration level is very similar to the solid solubility limit of Ga in Ge (4×10^{20} cm^{-3}, [148]). A similar correlation with the solid solubility limit was previously observed for the active concentration level of B in Ge [123, 152].

2.2.2 *Furnace Annealed Boron Junctions*

The goal of this section is twofold. Firstly, ion-implanted boron junctions in Ge and their behavior during subsequent furnace anneals are investigated. Secondly, the boron junctions are co-implanted with arsenic and fluor, thus creating samples resembling the integration of B junctions in a bulk Ge MOSFET flow.

2.2.2.1 Experimental Details

Boron junctions were fabricated using n-type, 200 mm diameter, (100)-oriented Si wafers on which a relaxed epitaxial Ge layer was grown (1.5 µm thick, threading dislocation density approx. 2×10^{7} cm^{-2}). All wafers (D02, D05 and D11 are discussed in this section) received phosphorus well implants, yielding an n-well with a doping concentration of 3×10^{17} cm^{-3}. Two wafers (D02 and D05) were implanted with As (80 keV, 5×10^{13} cm^{-2}, 15° tilt), followed by a F implant on D05 (4.3 keV, 1.6×10^{15} cm^{-2}, 7° tilt). After these I/I steps, the wafers were capped with a 10 nm thick protective SiO_2 layer. Subsequent annealing was performed in

Table 2.1 Overview of relevant quantities for the samples used in this study

Wafer number	D11	D02	D05
n-well (multiple P implants & anneal)	×	×	×
I/I: As 80 keV, 5×10^{13} cm^{-2} 15° tilt		×	×
I/I: F 4.3 keV, 1.6×10^{15} cm^{-2} 7° tilt			×
I/I: B 2.4 keV, 8×10^{14} cm^{-2} 7° tilt	×	×	×
SiO_2 deposition (10 nm)	×	×	×
RTA anneal 550 °C, 5 min, N_2	×	×	×
R_{sh} (4PP-tool)	–	failed	–
R_{sh} (VPS-technique)	–	83.5[a]	–
R_{sh} (μ4PP)	488.2	153.2	425.6

[a] R_{sh} value is too low due to probe penetration (measuring the substrate in parallel with the B-doped layer)

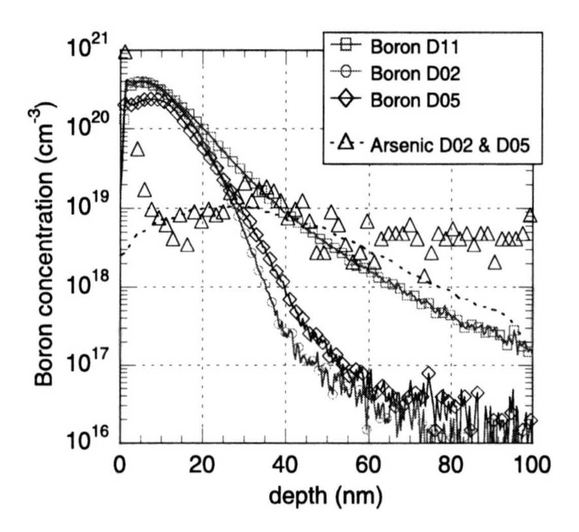

Fig. 2.6 Chemical concentration profiles as a function of depth for the boron (2.4 keV 8×10^{14} cm^{-2}), implanted in Ge and annealed at 550 °C for 5 min. Processing details can be found in Sect. 2.1

N_2 atmosphere in a *Heatpulse 610* rapid thermal annealing (RTA) system at 550 °C for 5 min. The chemical dopant concentration was studied by secondary-ion mass spectroscopy (SIMS), the electrical activation of B was analyzed by sheet resistance measurements with a conventional four-point probe (4PP) system, the variable probe spacing technique (VPS, [25]) and a micro-four-point probe system (μ4PP, [112]). The processing details of these samples are summarized in Table 2.1.

2.2.2.2 Physical Characterization

The chemical dopant concentration profiles in these wafers were analyzed using SIMS (Fig. 2.6). In wafer D11, which only received the B I/I, a pronounced tail is visible. This can be attributed to ion channeling during the implant and results in a rather deep junction. In wafer D02, the B profile is much shallower as a result of the preceding As implant. The implantation of As with a dose of 5×10^{13} cm^{-2} amorphizes the Ge to a depth of 37 nm (as measured using spectroscopic ellipsometry), thus eliminating ion-channeling during the subsequent B implant. As a result, the junction depth is reduced to 27 nm. Finally, in wafer D05, where F was implanted as well, considerable B dose-loss and slight in-diffusion can be observed. This reduces the total retained B dose and increases the junction depth again compared to sample D02 to $X_J = 30$ nm.

2.2.2.3 Electrical Characterization

On these B junctions, the sheet resistance R_{sh} was measured by conventional 4PP, the VPS technique and the μ4PP system. The first two techniques, while successful for the deeper Ga junctions in Sect. 2.2.1, did not deliver reliable results on the B samples (with and without the oxide cap on the sample). This is caused by the fact that these B junctions are even shallower and 4PP needles penetrate through the junction (contacting also the substrate). Consequently, the junction sheet resistance is measured in parallel with the substrate yielding a too low R_{sh}-value. Reliable measurements however could be obtained using the penetration-less μ4PP tool (oxide cap removed—see Table 2.1).

For wafer D11, R_{sh} is found to be 488.2 Ω/sq., corresponding to a calculated maximum active B concentration level of 1×10^{19} cm^{-3} (using the same methodology and mobility model as in Sect. 2.2.1.3). Similar low active concentration levels have been reported before for B implants into c-Ge and subsequent RTA annealing [123]. In order to reach a higher active boron concentration, Solid Phase Epitaxial Regrowth (SPER) is required: the full recrystallization of amorphous Ge has been reported to yield active B concentrations up to 6×10^{20} cm^{-3} [98], depending on anneal conditions. In wafer D02, the amorphous layer created by the As implant fully recrystallizes during the anneal. As such, a high active boron concentration level (4×10^{20} cm^{-3}) is obtained through SPER resulting in a low $R_{sh} = 153.2$ Ω/sq. Finally, in wafer D05, a higher R_{sh} is measured (425.6 Ω/sq.). This can be explained by the significant out-diffusion of B during the RTA anneal. As a result, the calculated maximum active concentration level drops to 1.8×10^{20} cm^{-3}.

2.2.3 *Conclusions*

In the previous section, gallium and boron were studied as possible p-type dopants in Ge. For Ga, a high active concentration (4.4×10^{20} cm^{-3}) was obtained without preceding Ge preamorphization (PAI) of the substrate. The low activation temperature (400 °C), combined with the absence of Ga-diffusion up to a temperature of 700 °C, make Ga junctions in crystalline Ge promising candidates for implementation in a High Performance, Short-Channel Ge technology. In the amorphous Ge phase, an increased Ga diffusivity and lower active concentration was observed following RTA-anneals above 400 °C. B junctions were fabricated showing similar high active concentrations (4×10^{20} cm^{-3}) for a junction depth as small as 27 nm. This however required a preamorphization of the Ge lattice prior to the B I/I such that B is efficiently incorporated through the mechanism of solid phase epitaxial regrowth into the Ge lattice during the RTA anneal. Finally, co-implanting with fluor caused significant B dose loss during the RTA anneal, yielding a 2× lower active concentration.

2.3 n-Type Junctions

2.3.1 Laser Annealed Arsenic Junctions

In contrast to their p-type counterparts [70, 98], RTA-annealed n-type junctions in Ge suffer from significant concentration-enhanced diffusion (using phosphorus or arsenic as dopant, [14, 36]). As such, the fabrication of ultrashallow n-type junctions requires limiting the activation anneal's thermal budget. At the same time, a higher temperature is required to achieve a high electrically active concentration. These two competing requirements have led to the use of ultrafast heat-treatment methods such as flash-assisted annealing (FLA, [51, 117, 132]) and millisecond laser annealing (LSA, [68, 150]). In this section, the feasibility of LSA to fabricate ultrashallow, low-resistive As junctions in Ge is studied. More specifically, the effects of the laser peak wafer temperature and the combination with a preamorphization implant (PAI) will be discussed.

2.3.1.1 Experimental Details

Arsenic junctions were fabricated using p-type, 300 mm diameter, (100)-oriented Si wafers on which a relaxed epitaxial Ge layer was grown (1.5 μm thick, threading dislocation density approx. 2×10^7 cm^{-2}). After growing a protective 2 nm GeO_2 layer, they received a boron n-well doping up to 3×10^{17} cm^{-3}, followed by a Ge PAI (20 keV, 2×10^{14} cm^{-2}) on selected samples, yielding an amorphized layer of 25 nm. Arsenic was implanted at an energy of 5 keV up to a dose of 5×10^{14} cm^{-2}. The samples then received millisecond laser annealing. The laser spot measured 1.1 cm × 75 μm and scans the wafer left-to-right at a speed of 75 mm/s (two consecutive scans). During the laser illumination, a wafer preheating is applied (250 °C) to reduce thermal stress arising from the localized laser heating. No absorber layer was deposited to assist in the laser anneal. Multiple regions (each measuring at least 5 × 5 cm) were illuminated, whereby the laser energy was varied in steps of 100 °C to reach peak wafer temperatures up to 900 °C (close to Ge's melting temperature

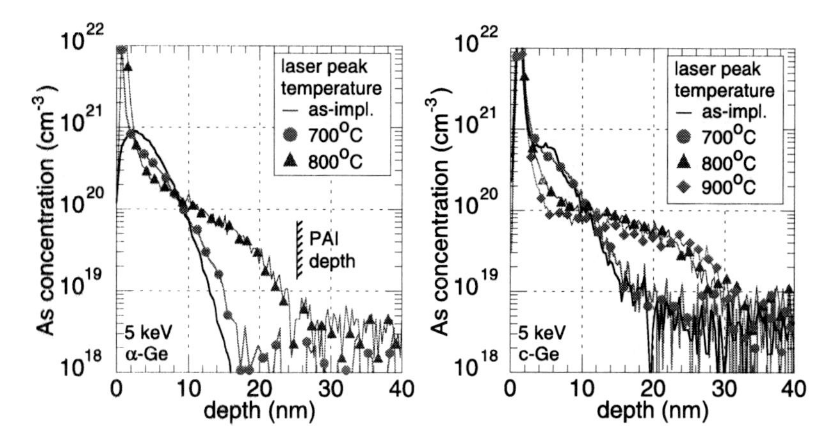

Fig. 2.7 Chemical concentration profiles as a function of depth for the As junctions (5 keV 5×10^{14} cm^{-2}), as-implanted in preamorphized (**a**) and crystalline (**b**) Ge and after LSA anneal

$T_{melt} = 937$ °C). This temperature staircase was calibrated by observing the onset of melting on the *c*-Ge wafers and assuming a linear dependency. Following this assumption, reported peak wafer temperatures should be considered an estimate. The chemical dopant concentrations was studied by secondary-ion mass spectroscopy (SIMS), the electrical activation of As was analyzed by sheet resistance measurements using the micro-four-point probe system (μ4PP, [112]). Additionally, Hall mobility measurements were also performed using the same system. An overview of the different samples and measured quantities is given in Table 2.2. The implant-induced damage and subsequent recrystallization were studied with transmission electron microscopy (TEM) and spectroscopic ellipsometry.

2.3.1.2 Physical Characterization

To investigate the diffusion behavior, the chemical As concentration was analyzed using SIMS. The resulting concentration profiles have a rather high background noise level (range mid-10^{18} cm^{-3}). This is due to mass interference of the ^{75}As with ^{74}Ge-^{1}H (^{74}Ge is the main Ge isotope). Figure 2.7(a) (α-Ge wafers) and Fig. 2.7(b) (*c*-Ge wafers) contain the resulting As concentration profiles for the laser annealed junctions for LSA wafer peak temperatures of 700, 800 and 900 °C. No significant As diffusion is observed up to a peak temperature of 700 °C, while considerable in-diffusion is present for samples annealed at 800 and 900 °C. The box-like profile after diffusion in the samples suggests that the concentration-enhanced diffusion mechanism (observed in RTA experiments, [125]) still applies. However, As diffusion seems to be rather independent of temperature in the range 800–900 °C, resulting in very similar As profiles for these annealing conditions. This enhanced diffusion leading to a box-like As profile is not observed on the α-Ge samples, where recrystallization occurs during LSA. A similar effect was seen in Si technology for

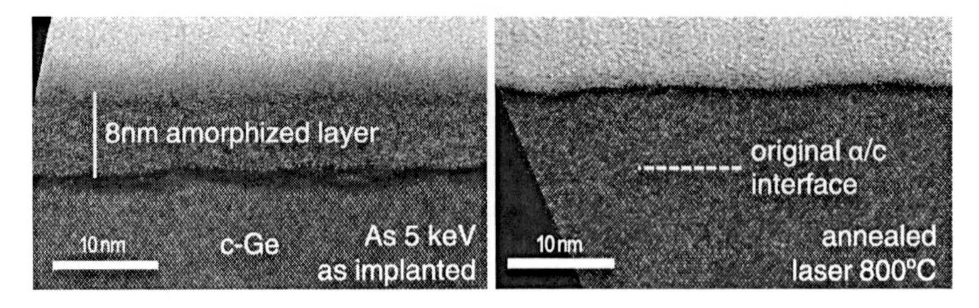

Fig. 2.8 Cross-sectional TEM of As-implanted *c*-Ge before (**a**) and after an 800 °C LSA anneal (**b**) showing full recrystallization

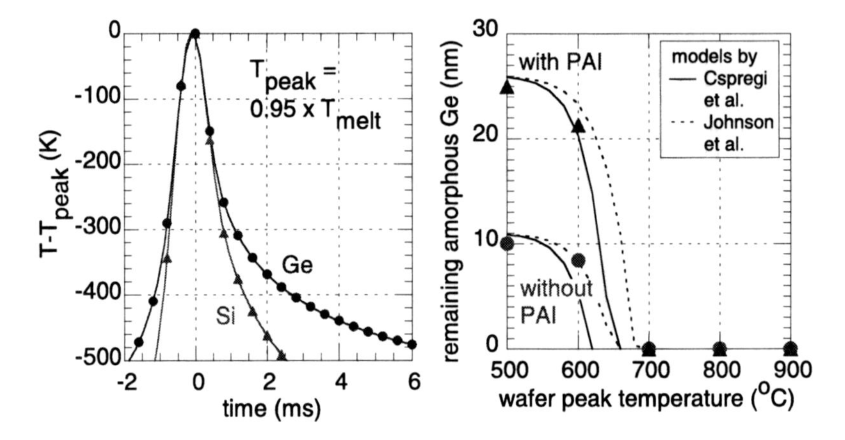

Fig. 2.9 (**a**) TCAD-simulated wafer surface temperature as a function of time for Si and Ge (peak temperature set to 95 % of T_{melt}). (**b**) Measured remaining thickness of the amorphous layer created by the ion implant after LSA anneal. The observed recrystallization is compared to literature using the TCAD-generated temperature profile

laser annealed As junctions, where PAI gives rise to a higher electrical activation [106]. Even though the same laser power was applied to the *c*-Ge and α-Ge samples (see Table 2.2), localized surface melting was observed on the 900 °C α-Ge sample, yielding a highly non-uniform junction, unsuited for further analysis. This observation may be caused by a decrease in surface reflectivity due to the amorphization, hence increasing the energy absorbed by the sample during LSA. Consequently, the estimated peak temperature for the α-Ge samples may be slightly higher than that of their *c*-Ge counterparts. Alternatively, this may be due to a difference in melting temperature between amorphized and crystalline Ge (as previously observed for Si in [143]).

The As implant results in amorphized Ge to a depth of 8 nm, as shown by the TEM micrographs in Fig. 2.8. Full recrystallization is observed after LSA at 800 °C. No residual crystal damage is observed at the original a/c interface. In contrast, recrystallization during LSA is known to be more cumbersome in Si, requiring addi-

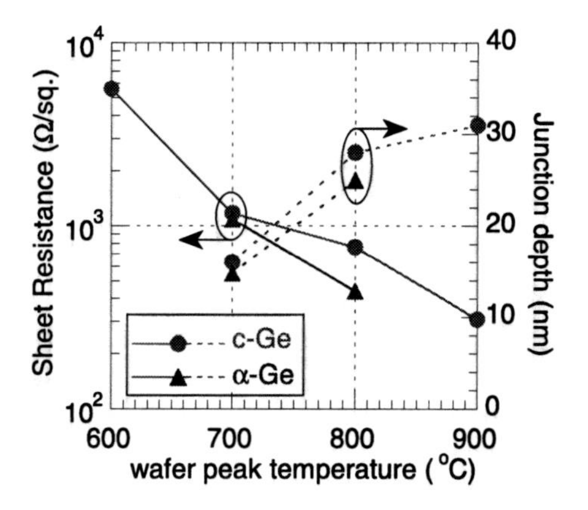

Fig. 2.10 Sheet resistance as measured with the μ4PP and junction depth for the laser annealed As junctions as a function of LSA peak temperature

tional treatments to cure remaining crystallographic defects [129]. To further investigate the recrystallization during LSA, a $T(t)$-profile was obtained using thermal TCAD simulations [128] to which the relevant material parameters for Ge were added (thermal conductivity [49], specific heat capacity [107]). For comparison, the same simulation was performed on Si substrates, using available default parameters. A typical $T(t)$-profile is plotted in Fig. 2.9(a). The LSA power was chosen to result in a peak wafer temperature $T_{peak} = 0.95 \times T_{melt}$. Note the striking similarity in the $T(t)$ profile in both materials, despite the difference in material parameters (the divergence from 2–6 ms is due to a different preheat temperature during LSA). This leads to the conclusion that the properties of the LSA tool (laser spot size, scanning speed, preheat temperature, etc.) are dominantly responsible for the shape of the temperature profile as a function of time.

The thickness of the remaining amorphous Ge as a function of T_{peak} was also measured with spectroscopic ellipsometry. Given the simulated $T(t)$ profiles, these can be compared to the expected recrystallization based on available Ge recrystallization models [29, 77] in Fig. 2.9(b). Firstly, reasonable agreement is obtained between the measurements and the regrowth models. Secondly, the amorphous layer is observed to be fully recrystallized after laser anneal with $T_{peak} = 700\,°\text{C}$, which is in line with the recrystallization models. Thirdly, whereas high P concentrations were shown to retard the regrowth substantially in [131], similar behavior could not be observed in the presence of As during our experiments. However, dedicated experiments would certainly be required to study this in-depth.

2.3.1.3 Electrical Characterization

The electrical activation of the arsenic junctions was studied through sheet resistance and Hall mobility measurements. Both of these were carried out using the μ4PP tool as described in [113], since the probe needles of a regular 4PP tool would definitely penetrate too deep into the Ge (given that 4PP already failed for deeper B junctions in Sect. 2.2.2) [26]. The measured and extracted quantities discussed in this section are listed in Table 2.2. The reported sheet resistances result

Table 2.2 Overview of relevant quantities for the samples used in this study

Sample number	Substrate type	Peak T (°C)	Laser energy (J/cm^{-2})	X_J (nm)	R_{sh} (Ω/sq.)
1	*c*-Ge	700	43.9	16	1173 ± 3 %
2	*c*-Ge	800	53.6	28	765 ± 2 %
3	*c*-Ge	900	63.4	31	308 ± 0.3 %
4	α-Ge	700	43.9	15	1089 ± 5 %
5	α-Ge	800	53.6	25	438 ± 3 %
6	α-Ge	900	63.4	*surface melting*	

Sample number	Hall μ (cm^2/Vs)	Hall electron density (cm^{-2})	Activation (%)	Calculated active concentration (cm^{-3})
1	80 ± 9 %	6.67×10^{13}	13	2.4×10^{19}[a]
2	96 ± 6 %	8.54×10^{13}	17	1.8×10^{19}
3	126 ± 5 %	1.61×10^{14}	32	5.0×10^{19}
4	76 ± 15 %	7.54×10^{13}	15	2.7×10^{19}[a]
5	111 ± 6 %	1.28×10^{14}	26	4.6×10^{19}
6	*no data—surface melting observed*			

[a]Uncertainties on the mobility model and especially on the SIMS close to the surface can cause large (±50 %) errors on the calculated activation level for the shallowest junctions

from multiple measurements on random locations for each sample and vary within 5 %, indicating that a uniform junction was formed, as in previous work [114]. The sheet resistance (R_{sh}) and the junction depth (X_J) as a function of peak wafer temperatures are plotted in Fig. 2.10. Firstly, in all investigated samples, a higher peak temperature is found to result in a deeper junction with a lower R_{sh}. Secondly, the α-Ge samples show lower R_{sh} in combination with a reduced X_J, indicating a higher electrical activation level. Note that X_J is extracted from the SIMS profile at a concentration of 10^{19} cm^{-3} (this is necessary because of the mid-10^{18} cm^{-3} noise-level on the SIMS).

From the SIMS analysis, R_{sh} and a concentration-dependent mobility model of As-doped germanium [44], an active concentration level was calculated. For the *c*-Ge samples, a maximum active concentration level of 5.0×10^{19} cm^{-3} was found for the 900 °C laser anneal. Comparing the α-Ge and *c*-Ge samples at 800 °C, the effect of the Ge PAI and the subsequent SPER can be seen: the maximum active As level increases from 1.8×10^{19} to 4.6×10^{19} cm^{-3}. This higher electrically active As concentration level can be explained by considering that As deactivation and diffusion in Ge are both known to be caused by the formation of mobile arsenic-vacancy (As-V) complexes [22]. As such, the formation of As-V complexes results in As de-activation and diffusion in the *c*-Ge samples, while As-V complexes are formed in lower concentrations in α-Ge samples due to the SPER. Another indication for this model is that the 900 °C *c*-Ge sample shows very little additional As diffusion with respect to the 800 °C c-Ge sample, consistent with theoretical calculations [22] predicting that As-V complexes are increasingly unstable at temperatures above 700–800 °C.

The electrically active As concentration of 5.0×10^{19} cm^{-3} in the 900 °C c-Ge sample is similar to the solid solubility limit of As in Ge at this temperature (i.e. 6×10^{19} cm^{-3} [134]), suggesting an efficient incorporation of As into the Ge lattice. The active As concentration also exceeds the level generally achieved using RTA annealing techniques [24].

The observed Hall mobility measurements (with the μ4PP tool) are also reported in Table 2.2. Observed electron mobility levels (ranging 76 to 126 cm^2/Vs) are lower than those reported in literature for similar electron concentrations (200 to 300 cm^2/Vs, [29]). At least to some extent, this is attributed to excess inactive As, acting as scattering sites.

2.3.2 Conclusions

In the previous section, arsenic was studied as possible n-type dopant in Ge. Ultra shallow arsenic junctions in Ge were fabricated using millisecond laser anneal (LSA) with peak wafer temperatures up to 900 °C. Significant in-diffusion was observed for samples annealed at 800 and 900 °C, yielding a box-like As profile suggesting a mechanism of concentration enhanced diffusion through the formation of arsenic-vacancy (As-V) complexes. A reduced formation of these complexes during LSA in α-Ge samples results in less diffusion and increased electrical activation. An electrical activation of 5.0×10^{19} cm^{-3} was achieved with the 900 °C laser anneal for $X_J = 31$ nm and $R_{sh} = 308\ \Omega/\text{sq}$. with $\mu_H = 126$ cm^2/Vs. Finally, LSA annealing is observed to result in full recrystallization of the amorphized Ge layer.

2.4 Benchmarking

In this section, junctions fabricated in this chapter are benchmarked with respect to results available in literature and the High Performance ITRS requirements for VLSI technology nodes [74, 75]. Additionally, the electrical activation requirements for ultra shallow junctions (USJ) in silicon and germanium will be discussed.

2.4.1 Electrical Activation Requirements

The maximum active concentration level is often used as a measure to compare different fabrication techniques for shallow junctions, since, unlike the sheet resistance R_{sh}, this quantity does not depend on the junction depth X_J itself. Although

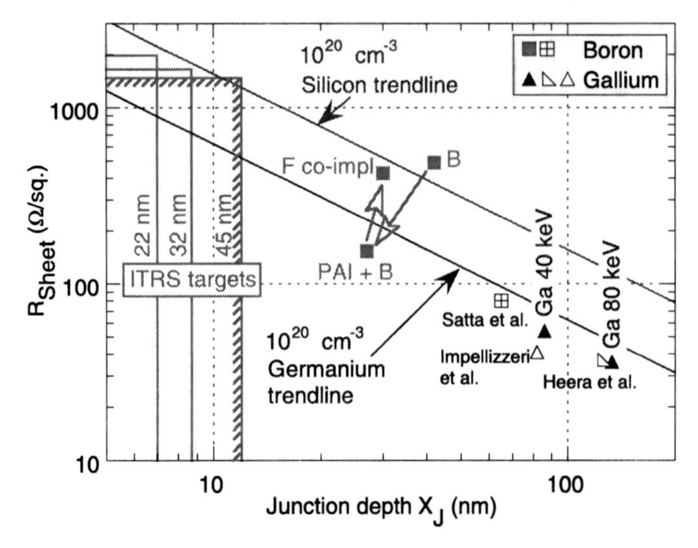

Fig. 2.11 p-Type junction sheet resistance as a function of junction depth for various junctions in Germanium. Trend lines for ideal box-like profiles are added for Si and Ge, along with ITRS targets for the 45, 32 and 22 nm nodes

useful, it remains a poor metric when comparing junctions in different materials. The reason for this is that R_{sh} at a given X_J (i.e. in a given technology) depends not only on the number of charge carriers, but also on the mobility. For an ideal box-like junction, the sheet resistance is given by:

$$R_{sh} = \frac{1}{q\mu(N_{ACT})N_{ACT}X_J} \tag{2.1}$$

Using Eq. (2.1), one can calculate that a p-type junction in Si with $N_{ACT} = 4 \times 10^{20}$ cm^{-3} ($\mu_{\mathrm{Si,h}}(N_{ACT} = 4 \times 10^{20}) = 24$ cm^2/Vs, [34]) has about the same sheet resistance as a Ge junction with a four times lower $N_{ACT} = 1 \times 10^{20}$ cm^{-3} because of the higher hole μ in this material ($\mu_{\mathrm{Ge,h}}(N_{ACT} = 1 \times 10^{20}) = 100$ cm^2/Vs, [50]). In turn, an n-type junction in $\mathrm{In_{0.53}Ga_{0.47}As}$ with an active concentration of only $N_{ACT} = 4 \times 10^{18}$ cm^{-3} has the same R_{sh} as an n-type Si junction, with 4×10^{20} cm^{-3} ($\mu_{\mathrm{Si,e}}(N_{ACT} = 4 \times 10^{20}) = 30$ cm^2/Vs, [86]) because of the high electron mobility in this material ($\mu_{\mathrm{In_{0.53}Ga_{0.47}As,e}}(N_{ACT} = 4 \times 10^{18}) = 3000$ cm^2/Vs, [110]). These considerations, among others, make the quantity N_{ACT}, while useful for other purposes, less suited for a benchmarking exercise focussed on the integration of USJ in VLSI logic when comparing different semiconductors.

2.4.2 Benchmarking of USJ in Germanium

The benchmarking of junctions in Ge will be performed using R_{sh} and X_J. R_{sh} is measured directly using a sheet resistance probing technique such as 4PP, VPS or

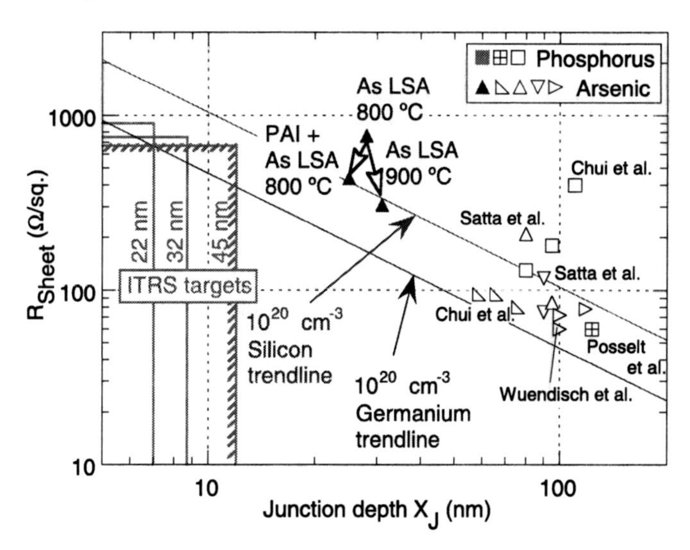

Fig. 2.12 n-Type junction sheet resistance as a function of junction depth for various junctions in Germanium. Trend lines for ideal box-like profiles are added for Si and Ge, along with ITRS targets for the 45, 32 and 22 nm nodes

μ4PP). X_J is taken as either X_J from SIMS-measurements at a concentration level of 4×10^{18} cm^{-3} or the reported metallurgical X_J (whichever one is smaller). This limits the comparison with literature to those references that actually report R_{sh} and X_J or allow their direct extraction (e.g. from the figures). Some authors have used the Spreading Resistance Probing technique (SRP) to characterize dopant activation. Still, calculating R_{sh} from an SRP profile can result in rather large error bars on the result. Consequently, those R_{sh} values will not be included.

2.4.2.1 p-Type Junctions

For p-type junctions, this results in the R_{sh}–X_J plot in Fig. 2.11. The graph includes literature references for B-junctions [125] and for Ga-junctions [51, 70] in Ge, as well as our own results which were discussed in Sects. 2.2.2 (Boron) and 2.2.1 (Gallium). In addition to these data points, the calculated sheet resistance values were included for an ideal box-like p-type junction in Si and Ge ($N_{ACT} = 1 \times 10^{20}$ cm^{-3}) with varying X_J. Finally, the ITRS targets for the 45, 32 and 22 nm technology nodes [74, 75] were added. While the majority of the reported Ge junctions are still a lot deeper than the targets, their electrical activation is equivalent to ideal junctions with N_{ACT} well above 10^{20} cm^{-3}. This graph also underlines the importance of the results obtained in Sect. 2.2.2, where a 27 nm junction was fabricated with a high concentration of electrically active B using SPER and RTA-anneal. If this trend can be maintained and similar active concentration levels can be achieved in even shallower junctions, the sheet resistance of these p-type junctions would be 3–4 times lower than the corresponding ITRS targets.

2.4.2.2 n-Type Junctions

For n-type junctions, a similar R_{sh}–X_J plot was made (Fig. 2.12). This graph includes literature references for p-junctions [24, 117] and for As-junctions [24, 124, 125, 156] in Ge, as well as our own results for As junctions (which were discussed in Sect. 2.3.1). Calculated sheet resistance values were included for an ideal n-type box-like junction in Si and Ge ($N_{ACT} = 10^{20}$ cm^{-3}) with varying X_J. Finally, the ITRS targets for the 45, 32 and 22 nm technology nodes [74, 75] were added. Note that the required R_{sh} values are lower, since the electron mobility exceeds the hole mobility, giving rise to a lower R_{sh} for a given N_{ACT} (see Eq. (2.1)). The n-type junctions found in literature are all deeper than 60 nm. Their sheet resistance is higher than that of an ideal n-type box-like profile ($N_{ACT} = 10^{20}$ cm^{-3}). Using LSA annealing, As junctions were successfully scaled to the $X_J = 20$–30 nm range (Sect. 2.3.1). Nevertheless, it is clear that scaling X_J while maintaining (or preferably improving) the electrical activation is required to fulfill ITRS targets. Unlike p-type dopants in Ge, the higher diffusivity and lower electrical activation of P and As complicates n-type USJ fabrication in Ge.

2.4.3 *Conclusions*

In the previous section, the electrical activation requirements for ultra shallow junctions (USJ) were discussed. It was found that the electrically active doping concentration requirements in materials which possess a higher carrier mobility than silicon can be much lower (keeping R_{sh} and X_J constant). Taking ideal Si junctions with an active carrier concentration of 4×10^{20} cm^{-3} as a reference, similar R_{sh} values can be obtained in Ge with a 4× lower N_{ACT}, and in $In_{0.53}Ga_{0.47}As$ with a 100× lower N_{ACT}. As such, high mobility materials offer opportunities to reduce parasitic series resistance in MOSFETs, increasing drive current for scaled devices.

Using R_{sh}–X_J plots, the Ge junctions fabricated in this chapter were benchmarked against recent literature results and the ITRS targets for the upcoming technology nodes:

- *For p-type junctions*, a rather high electrically active dopant concentration level was seen for B and Ga junctions (up to 6×10^{20} cm^{-3}, [70]). This level is maintained in shallow junctions (down to $X_J = 27$ nm, 4×10^{20} cm^{-3}). If this trend can be upheld, the sheet resistance of p-type junctions in Ge would be 3–4 times lower than the ITRS targets for the corresponding technology nodes.
- *For n-type junctions*, a lower electrically active dopant concentration level is observed for P and As junctions than for the p-type junctions (10^{19} cm^{-3}-range, [24]), despite the use of more advanced annealing techniques such as FLA and LSA. While an active concentration level of 5×10^{19} cm^{-3} has now been achieved in shallow junctions ($X_J = 25$–31 nm), further improvements are required to meet the n-type ITRS targets.

2.5 Summary and Conclusions

The goal of this chapter was to investigate the fabrication of (shallow) junctions in Ge. To this end, gallium and boron were considered as possible p-type dopants while arsenic was used to fabricate n-type junctions:

Gallium was implanted in crystalline and preamorphized Ge. After a 60 s, 550 °C RTA, a high electrically active dopant concentration level of 4.4×10^{20} cm^{-3} was observed. This study also revealed an increased Ga diffusivity in the amorphous Ge phase. Cross-sectional TEM analysis showed that the recrystallization (SPER) of the amorphous Ge layer is about 3× slower in samples which received a deep pre-amorphization implant (PAI).

Boron junctions were also fabricated showing similar high electrically active dopant concentration levels (4×10^{20} cm^{-3}). The lower implant energy used in this study allowed reducing the junction depth down to 27 nm. In order to achieve an efficient incorporation of Boron into the Ge lattice, a PAI is required, combined with SPER during the RTA anneal. Co-implanting B junctions with F was shown to degrade junction properties: increased B diffusion and resulting dose loss during the RTA anneal yields a 2× lower electrically active concentration level.

Arsenic was studied as an n-type dopant in Ge. Millisecond laser annealing (LSA) was used to activate the implanted arsenic in an attempt to reduce the concentration-enhanced diffusion and resulting dopant deactivation, commonly observed with n-type dopants in Ge. Significant diffusion was however still present in samples annealed at 800 and 900 °C. Furthermore, the box-like profile of the resulting As junctions suggests that the observed diffusion and deactivation during LSA still occurs through the same mechanism of mobile Arsenic-Vacancy complexes. Despite these issues, an electrically active concentration level of 5.0×10^{19} cm^{-3} was achieved with a 900 °C LSA for $X_J = 31$ nm and $\mu_H = 126$ cm^2/Vs following As I/I into c-Ge wafers. Cross-sectional TEM analysis showed full, defect-free recrystallization of the amorphized Ge layer during LSA provided the wafer reaches a peak temperature of 700 °C.

In order to compare the results obtained to the existing Si literature results, electrical activation requirements for ultra shallow junctions (USJ) were discussed. Calculations showed that the required electrically active doping concentration in high-mobility materials can be significantly lower than that in silicon. Taking an ideal Si junction with an active carrier concentration of 4×10^{20} cm^{-3} as a reference, simi lar R_{sh} values can be obtained in Ge with a 4× lower N_{ACT}, and in $In_{0.53}Ga_{0.47}As$ with a 100× lower N_{ACT}. As such, high mobility materials offer opportunities to reduce parasitic series resistance in MOSFETs, increasing drive current for scaled devices.

Finally, the Ge junctions fabricated in this chapter were benchmarked against recent literature results and the high-performance ITRS targets for the upcoming technology nodes. p-Type junctions using B and Ga in Ge were shown to combine high

N_{ACT} with diffusionless behavior under certain conditions. Consequently, while junction depths were still deeper than ITRS targets, it seems highly likely that junction depths can be reduced further by simply reducing the I/I energy. Provided the current N_{ACT} can be maintained, sheet resistance of B and Ga p-type junctions in Ge would be 3–4 times lower than the targets for the upcoming technology nodes (45, 32 and 22 nm). In contrast, a lower N_{ACT} was observed for P and As junctions in Ge, in combination with significant concentration-enhanced diffusion. So far, even more advanced annealing techniques such as FLA and LSA have failed to produce active dopant concentration levels matching those achieved with p-type dopants in combination with a sufficiently low X_J. Even though, using LSA, active concentration levels close to the ITRS requirements were obtained, further improvements are required to meet n-type junction targets in Ge.

Appendix

A.1 Thermal Laser Anneal—sprocess Simulation Parameters

These are the parameters used for the thermal laser annealing simulations in this work. The syntax below corresponds to the 2010.03 release of sprocess [128]. Default Si parameters are repeated as well for reference.

```
SILICON
Potential ni [expr pow([simGetDouble Diffuse tempK],1.5)
      *[Arr 2.16e16 [expr 0.36-5.12e-8*pow([simGetDouble
      Diffuse tempK],2)] ]]
Absorptivity  0.5*(1.9e-20*(Temperature ^ 1.5)
      *(1.0e15 + [pdbDelayDouble Silicon Potential ni])) + 900.0
Emissivity  1.0
SpecificHeatCapacity  1176 + Temperature
      * (1.3e-4*Temperature - 0.252) - 1.19e5 / Temperature
ThermalConductivity  0.01 * (-73.85 + Temperature
      * (-1.36e-5*Temperature + 5.72e-2) + 6.21e4 / Temperature)
```

```
GERMANIUM
 ## absorptivity
Potential ni [expr pow([simGetDouble Diffuse tempK],1.5)
      *[Arr 2.16e16 [expr 0.36-5.12e-8*pow([simGetDouble
      Diffuse tempK],2)] ]]
Absorptivity  0.5*(1.9e-20*(Temperature ^ 1.5)
      *(1.0e15 + [pdbDelayDouble Silicon Potential ni])) + 900.0
Emissivity  0.1\
SpecificHeatCapacity  347.6 + Temperature
      * (1.45e-5*Temperature - 0.00995) - 13921 / Temperature
ThermalConductivity  0.0169 * (-38.715 + Temperature
      * (-2.94e-5*Temperature + 5.90e-2) + 1.76e4 / Temperature)
```

Chapter 3
TCAD Simulation and Modeling of Ion Implants in Germanium

In this chapter, a TCAD process simulator is used to model dopant implants in germanium. It is then applied to design ion implant steps for a scaled $L_G = 70$ nm germanium pMOSFET technology.

3.1 Introduction

Ion implantation is one of the most widely used processing techniques to introduce impurity atoms into semiconductor materials. Consequently, detailed modeling of ion implantation distributions is a requirement for accurate TCADprocess simulations. The Monte Carlo (MC) simulation method is often used to compute the distribution of implanted ions and the resulting implantation damage. Following years of development, MC simulators have been calibrated to include many combinations of incident ions and substrate atoms (e.g. the SRIM simulator, [161]). However, many only consider amorphous substrate materials. As a result, they cannot capture effects caused by the crystallographic nature of common semiconductor substrates, such as ion channeling. To tackle this issue, MC simulators have been developed and implemented for Si substrates in commercial TCAD software [128] which include the crystallinity of the substrate material, such as crystalTRIM, [116] and TaurusMC, [146]). These simulators lack the required material parameters and calibration to allow detailed modeling of ion implantation into Ge substrates. As such, the first goal of this chapter will be to calibrate one of the MC simulators for ion implants into Ge, including crystallographic effects.

Secondly, although a calibrated Monte Carlo simulator would allow for reliable ion implant simulations, these will inevitably be computationally intensive: statistical noise on ion implant profiles can only be reduced by increasing the number of simulated pseudo-particles, a limitation inherent to the Monte Carlo method. To address this problem, an analytical model will be proposed, describing the ion implant profiles with analytical distribution functions. This approach allows calculation of low-noise ion implant profiles in Ge without the need for time consuming MC simulations.

G. Hellings, K. De Meyer, *High Mobility and Quantum Well Transistors*,
Springer Series in Advanced Microelectronics 42, DOI 10.1007/978-94-007-6340-1_3,

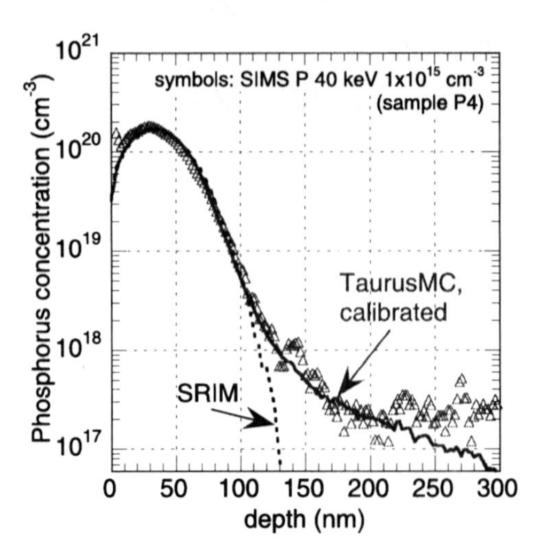

Fig. 3.1 SIMS measurements (*symbols*) and simulated P profiles using the SRIM Monte Carlo simulator [161] and the calibrated TaurusMC

Finally, the calibrated Monte Carlo simulator will be used to design the various ion implant steps required for a scaled $L_G = 70$ nm Ge pFET technology, imitating optimized doping profiles of a 65 nm Si pFET fabrication flow. Fabrication of Ge pFETs with L_G down to 70 nm allows comparison of different implant conditions for the extensions and the halos and benchmarking of these devices versus existing bulk Ge references in literature and the ITRS requirements for the corresponding technology node. Note that only Ge pFETs are considered in this chapter, since essential elements for the fabrication of scaled Ge nFETs need further investigation (e.g. shallow, low-resistive n-type junctions and nFET gate stack passivation).

3.2 Ion Implant into Germanium—Monte Carlo Simulations

The goal of this section is to discuss the calibration of the TaurusMC Monte Carlo TCAD tool for simulating ion implants into Germanium substrates. As mentioned above, accurate simulations of the as-implanted dopant profile in a crystalline solid require the inclusion of ion channeling. One example of this necessity is given in Fig. 3.1, which compares a measured (SIMS) as-implanted phosphorus profile in Ge with two simulated ones. The first profile was obtained using the Stopping Range of Ions in Matter (SRIM) simulator [161]. The main limitation of this simulator is that it considers only structurally isotropic substrates. While this approach yields accurate profiles for implants into amorphous material, it fails to capture important effects in crystalline substrates. Indeed, the crystallinity of the target material introduces an important anisotropy: implanted ions can travel more easily in certain directions, an effect called ion channeling. As such, the SRIM simulator fails to reproduce the deeper part of the implant, where ion-channeling is observed (at depths exceeding 100 nm in Fig. 3.1). The second simulated profile was obtained using the

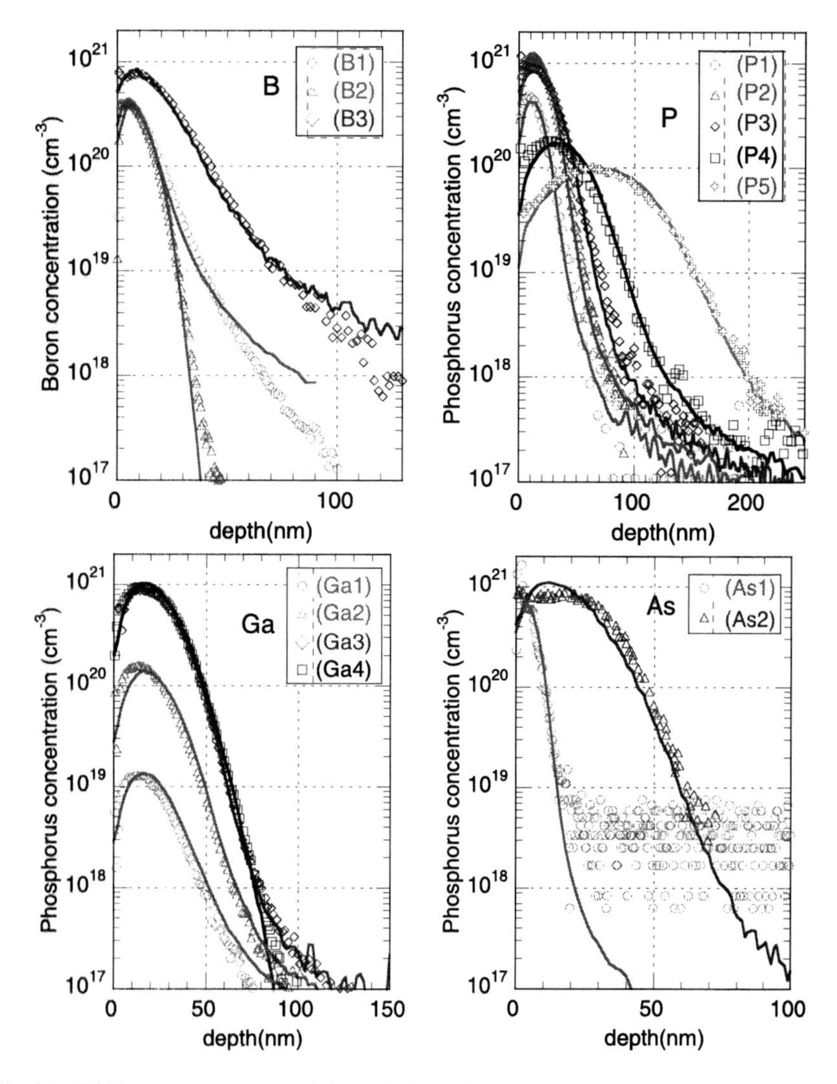

Fig. 3.2 SIMS measurements (*symbols*) and Monte Carlo simulations of I/I profiles into Ge, for various dopants, energies, doses. Sample details are in Table 3.1

calibrated Monte Carlo simulator TaurusMC, which also takes into consideration the crystallinity of the target substrate. Using this simulator, the ion channeling-related portion of the as-implanted profile can be reproduced.

At the beginning of an ion implantation step the crystallographic channels inside the Ge substrate are intact, allowing ions to travel more easily in these directions (ion channeling). However, as displaced lattice atoms clutter these crystallographic channels, this ion channeling process becomes less efficient. This is known as damage de-channeling [128]. Eventually, as even more ions are implanted, these crystallo-

Table 3.1 Overview of the samples used to calibrate the TaurusMC Monte Carlo simulator for I/I into Ge substrates

Sample ID	Reference	Species	Energy (keV)	Dose (cm^{-2})	Tilt-twist (°)	Cap layer (nm)	SIMS (Fig. 2.1)	Amorphous Ge thickness after I/I			
								TEM (nm)	SE (nm)	MC simulated (nm)	delta (%)
B1	Table 2.1 (D11)	B	2.4	8×10^{14}	7–27	none	×	no data			
B2[2]	Table 2.1 (D02)	B	2.4	8×10^{14}	7–27	none	×	no data			
B3	[125]	B	6	3×10^{15}	7–0	10	×	0		0	+0 %
P1		P	12	1×10^{15}	7–0	none	×	no data			
P2	[124]	P	15	3×10^{15}	7–0	none	×	no data			
P3	[125]	P	25	3×10^{15}	7–0	10	×	50		47	−6 %
P4		P	40	1×10^{15}	7–0	none	×	no data			
P5	[137][a]	P	80	1×10^{15}	7–0	none	×	no data			
Ga1	[55]	Ga	40	4×10^{13}	7–0	none	×	no data			
Ga2	[55]	Ga	40	4×10^{14}	7–0	none	×	no data			
Ga3	Section 2.2.1	Ga	40	3×10^{15}	7–0	10	×	58	59	54	−7.7 %
Ga4[b]	Section 2.2.1	Ga	40	3×10^{15}	7–0	none	×	n/a			
Ga5	[55]	Ga	80	3×10^{15}	7–0	none		93		92	−1.1 %
Ga6	[72]	Ga	150	5×10^{13}	7–0	none		108		80	−25.9 %
Ga7	[72]	Ga	150	1×10^{15}	7–0	none		137		138	+0.8 %
Ge1	Section 2.3.1	Ge	20	2×10^{14}	0–0	2			25	20	−20 %
Ge2	Section 2.2.1	Ge	200	1×10^{15}	0–0	none	190		172		−9.5 %
As1	Section 2.3.1	As	5	5×10^{14}	0–0	2	×	8	10	8.5	−5.6 %
As2	[125]	As	50	4×10^{15}	7–0	10	×	50		51.5	+3 %
As3		As	80	5×10^{13}	15–0	11.6			37.3	42	+12.6 %

[a] SIMS depth scale adjusted slightly

[b] Additional ion implants were performed on this sample (prior to the one mentioned)

graphic channels are destroyed completely: the ion bombardment causes complete amorphization of the substrate (to a certain depth), preventing further ion channeling. As a result of this dynamic process, the shape of ion implant profiles into crystalline materials can be heavily dependent on the implanted dose. Low-dose implants will result in a relatively larger channeling tail, while in high-dose implants, most ions will be implanted when the ion channels are already destroyed, yielding a relatively small ion channeling contribution.

A reliable calibration of TaurusMC for implantation of common dopants into crystalline Ge is not generally available. A first set of required parameters contains basic information about the Ge crystal lattice. These physical parameters are generally known and can readily be implemented for any substrate (e.g. atomic mass, mass density, lattice constant, crystal structure, ...). A second set of parameters controls nuclear and electronic stopping of the implanted ions, as they travel through the substrate as well as the damage accumulation in the crystal lattice during the ion bombardment simulation. This second set of parameters is obtained through numerical optimization, aiming for good fitting of the MC simulated profiles to a vast database of experimental SIMS profiles.

In order to calibrate the TaurusMC simulator, the second set of parameters was adjusted, starting from the Si defaults, until good fits were obtained when comparing experimental as-implanted profiles in Ge to their simulated equivalent. The resulting MC simulations and SIMS profiles are plotted in Fig. 3.2(a)–(d) for boron, phosphorus, gallium and arsenic I/I into Ge respectively. Note that the parameters for the I/I of Ga, Ge and As were changed together, as these three elements have a similar atomic mass. The sample processing details are listed in Table 3.1. On some of the samples, a capping layer (mostly SiO_2) was deposited prior to the actual implant (e.g. P3 and B3). Also, some samples received more than one I/I whereby crystallographic damage caused by the first implant can have a profound influence on the distribution of the second implant (e.g. B1 and B2). These effects were included in the MC simulations. Good agreement was obtained between the MC simulations and the SIMS data for P, Ga and As. A modest deviation is observed in the tail portion of the B profiles in Fig. 3.2(a), although the MC simulator is able to reproduce the main features of the profile.

As a fitting to SIMS profiles is still an indirect way to check the damage evolution in the sample during the ion implant and susceptible to measurement noise in the channeling tail, a more direct approach was taken to measure and calibrate the damage evolution in the Ge crystal lattice. For this, the thickness of the amorphous Ge layer created during I/I was measured on various samples using Cross-sectional Transmission Electron Microscopy (TEM) or Spectroscopic Ellipsometry (SE). These measured thicknesses were then compared to the MC simulator's predictions. As can be seen in Table 3.1, the difference between both is generally small, with typical deviations in the range of 0 to 10 %. The adjusted parameters for the TaurusMC simulator are listed in appendix to this chapter and should allow to obtain as-implanted profiles for common dopants in crystallographic Ge with reasonable accuracy.

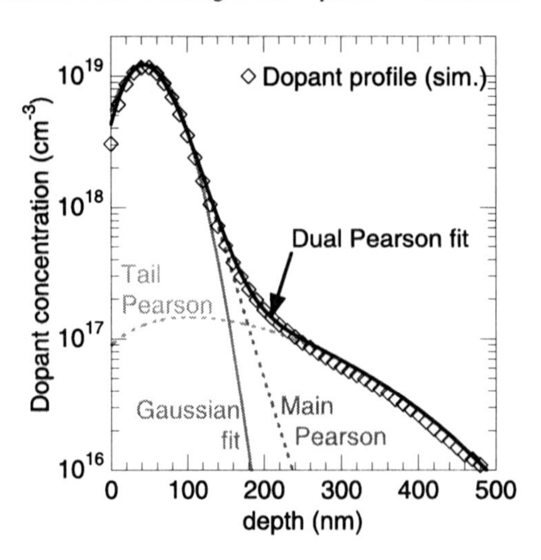

Fig. 3.3 Typical implant profile into a crystalline substrate (*symbols*) and analytical fits using various distributions: Gaussian, Pearson and Dual Pearson

3.3 Ion Implant into Germanium—Analytical Description

3.3.1 Dual Pearson Distribution Functions

Although the calibrated TaurusMC simulator produces good fits to experimental data, it suffers from the fact that the statistical noise inherent to Monte Carlo simulations can only be reduced by increasing the number of simulated pseudo-particles. Combined with the fact that dopant concentration profiles typically span several orders of magnitude, this has a detrimental impact on the required computational power to achieve smooth profiles. This especially limits the applicability of MC simulations when simulating complex 3-dimensional structures.

To address this problem, ion implant profiles have been described using analytical functions such as the Gaussian distribution. A library of calibration tables, containing the parameters of these analytical functions (as a function of species, energy, dose, tilt angle, cap layer thickness etc.) is constructed. This library then allows calculating ion implant profiles with limited computational power using interpolation.

Distribution functions most commonly used for this purpose are the simple Gaussian and the Pearson family of distribution functions of which the Gaussian curve is a member [5]. The tail resulting from ion channeling is often included by use of a second Pearson curve: the sum of the main Pearson and the tail Pearson then represents the entire implantation profile. This approach is illustrated in Fig. 3.3: A typical as-implanted profile from the MC simulator is fitted with a simple Gaussian, a Pearson and a dual Pearson curve. The Gaussian nor the single Pearson can fit the entire profile including the ion-channeling tail. The dual Pearson curve is able to fit the entire profile. Others have constructed different analytical distributions, attempting to capture more of the physics in the analytical model, using e.g. tail functions [135, 137]. Still, the dual Pearson curves are the dominant distribution functions used in commercial TCAD software [128].

The Pearson family of distribution curves contains 12 separately identifiable types and results from solving the differential equation [5]:

$$\frac{df(y)}{dy} = \frac{(x-a)f(y)}{b_0 + b_1 y + b_2 y^2} \tag{3.1}$$

The four parameters of Eq. (3.1) are related to the four moments of the Pearson distribution function: Range R_p, straggle σ_p, skewness γ_p and traditional kurtosis β_p. A dual Pearson curve thus requires 10 parameters: four moments and one normalization factor for each Pearson. Before discussing the analytical model, two remarks are made:

- In this study no restrictions were imposed on the type of Pearson curve used for fitting the ion implant curves. Often specific types of Pearson curves are used to describe ion implant profiles (e.g. the type IV Pearson was proposed by [67]). However as [5] has shown, many other types are also well suited for modeling of ion implant profiles. Therefore, in this work the above-mentioned differential equation (Eq. (3.1)) was solved using the four moments as parameters. The resulting analytic expression is a general solution for a Pearson curve as a function of its four moments: R_p, σ_p, γ_p and β_p.
- Various studies calculate the Pearson moments directly from an ion implant profile using the common statistical definition for the moments. However, Pearson moments calculated directly from the experimental or simulated profiles in this manner will not be sufficiently close to the real moments. The reason for this is that these are inevitably calculated on semi-infinite profiles (covering the range from the boundary presented by the semiconductor surface to a certain maximum analysis depth). Instead, the moments required should be calculated on the whole range of the distribution curve [5, 136]. Therefore, in this study, a numerical least-squares algorithm is used to fit each profile with a dual Pearson curve rather than calculating the moments directly, avoiding this problem.

3.3.2 Analytical Model

Monte Carlo ion implant simulations of P, Ga and As were performed over an energy range of 15–180 keV (P) or 30–360 keV (Ga and As) and for doses ranging from 10^{12} to 10^{16} cm^{-2} at 7° tilt. This range covers the implant conditions generally used in the typical Ge pMOSFETs [33] as well as lower energy conditions which can be of use to future development of both n- and pMOSFETs. Based on these profiles, an analytical model was constructed, allowing to fit the I/I profiles over this entire dose and energy range using dual Pearson curves. This model and its parameters are contained in Table 3.2. Most of the Pearson moments were fitted using a second order polynomial and can be considered a function of energy only. For the main Pearson curve, only the skewness is a function of both implant energy and dose, an effect linked to ion channeling. For the tail Pearson, the skewness and kurtosis

Table 3.2 Overview of the parameterized analytical model for ion implants into crystalline substrate

	Main Pearson	Tail Pearson
R_p (nm)	$p_0 + p_1 E + p_2 E^2$	$p_{14} R_p$
σ_p (nm)	$p_3 + p_4 E + p_5 E^2$	$p_{15} \sigma_p$
γ_p (–)	$p_6 + p_7 E + p_8 E^2 + p_9 (1 - (\frac{D}{D+p_{10}})^3)^3$	0
β_p (–)	$p_{11} + p_{12} E + p_{13} E^2$	3.36
dose (cm^{-2})	$D - D_{tail}$	$\begin{cases} D_{tail} = (1/d_a + 1/d_b)^{-1} \\ d_a = p_{16} D \\ d_b = (\frac{D}{10^{13}})^{p_{17}} p_{18} \end{cases}$

were fixed for numerical stability. Finally, the fraction of the implanted dose that is attributed to the tail Pearson (channeling dose) is modeled considering two regimes:

- For low dose implants, the fraction of ions that are contained in the channeling tail is constant as a function of the implanted dose. This can be explained by considering that for low implanted doses, the crystallographic channels inside the substrate remain intact during the entire I/I step. As such, every implanted ion has an equal chance of ending up in the channeling tail (d_a in Table 3.2, low-dose limit in Fig. 3.4).
- For high-dose implants, the absolute number of ions that are contained in the channeling tail is rather constant. This can be explained by considering that for high implanted doses, the crystallographic channels get destroyed at one point during the I/I. Any ions implanted beyond this point will not end up in the channeling tail, giving rise to an almost constant number of ions in the channeling tail. (d_b in Table 3.2, high-dose limit in Fig. 3.4).
- Finally, these two regimes are combined in final formula for the channeling dose (D_{tail} in Table 3.2) with a smooth transition.

The parameters of the resulting analytical model were obtained using a numerical optimization algorithm minimizing the total fitting error between the Monte Carlo simulated I/I profiles and the analytical profiles for P, As and Ga. The resulting values can be found in Table 3.5 in Appendix A.2. Boron was omitted here as a species, since the fit between the MC simulations and measured SIMS data was of lesser quality than for the other dopants. Note that the parameters p_0–p_{18} should purely be considered as empirical fitting parameters.

3.3.3 *Practical Applications*

A first practical application of this analytical model is obviously that it allows to quickly predict an ion implant profile. As an example of this, Fig. 3.5 contains three measured SIMS profiles (various Ga and P implants), the simulated profiles using

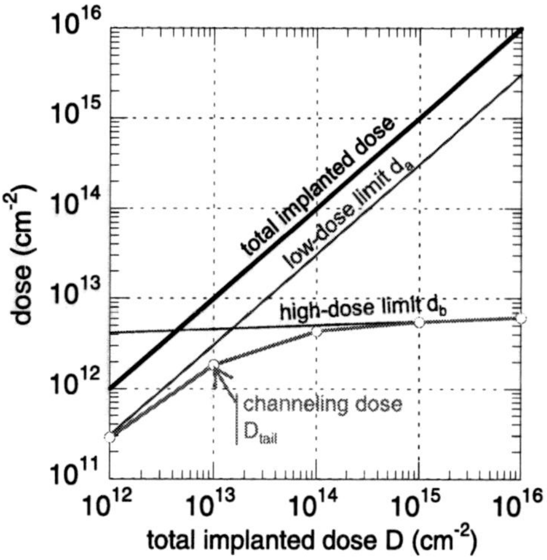

Fig. 3.4 Channeling dose dependence as a function of total implanted dose for phosphorus implants into crystalline germanium showing the combination of the low-dose limit d_a (channeling dose proportional to total implanted dose and the high-dose limit d_b (channeling dose nearly constant)

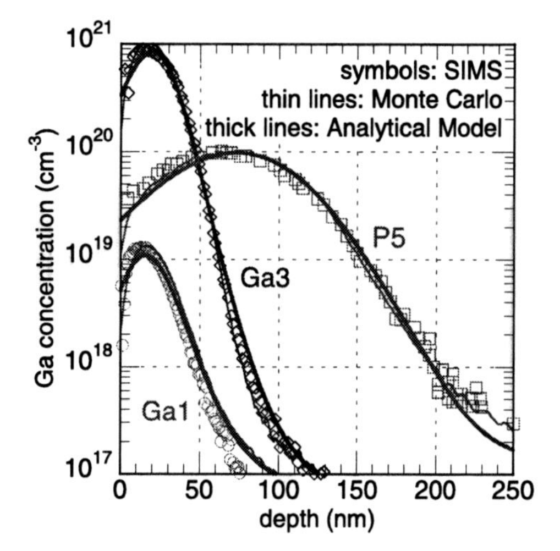

Fig. 3.5 Doping profiles of Ga and P in Ge as obtained from SIMS measurements (*symbols*), the calibrated Monte Carlo simulator (*thin lines*) and the analytical model (*thick lines*)

the calibrated MC simulator and the calculated profiles using the analytical model. For each of these conditions, the difference between all three are rather small. Note that the energy of these conditions was not used to obtain the model parameters. The advantage of using the analytical model becomes clear when comparing the time required to obtain such 1-dimensional profiles. While the MC simulator took 6 minutes (P5 profile), the analytical profile was calculated in less than 1 second. Additionally, the MC simulated profiles have clearly visible statistical noise (e.g. in the 10^{17} cm^{-3} range for the P5 profile). This performance gap increases even more when comparing simulations in 2D or 3D (e.g. process simulations of a typical transistor).

A second practical application is that this analytical model can be used to construct implantation tables for I/I of common dopants in germanium. Many modern process simulators use such calibration tables containing (dual) Pearson moments as a function of energy, dose, tilt angle, and other parameters. With the analytical model, such tables can be constructed containing many profiles (i.e. many energy/dose combinations) starting from a limited set of MC simulations.

3.3.4 Conclusions

In the previous sections, the Taurus Monte Carlo simulator was calibrated for ion implants of boron, phosphorus, gallium, germanium and arsenic into Germanium based on experimental SIMS profiles. Additionally, the thickness of the amorphous layer induced by the I/I was used as a second calibration using TEM and SE measurements. As such, accurate simulations of as-implanted dopant profiles can now be performed, taking into account such crystal lattice effects as ion-channeling in Germanium. An analytical model was proposed based on the description of as-implanted dopant profiles using dual Pearson distribution curves (P, Ga, As). This model spans an energy/dose range which should cover most of the I/I conditions relevant for short channel Ge p- and nMOSFET development. This analytical description allows calculation of low-noise I/I profiles in Ge without the need for time consuming MC simulations.

3.4 Application to a 70 nm Bulk Ge pFET Technology

In this section, high performance Ge pMOSFETs will be presented with physical gate lengths down to 70 nm. The various implant steps were designed using the calibrated MC implant simulator to mimic the doping profiles in an optimized 65 nm Si pFET fabrication flow. On top of this, different extension and halo implant conditions are compared. Finally, the 70 nm Ge pFET device is benchmarked against existing bulk Ge references in literature and against ITRS requirements for the corresponding technology node.

3.4.1 Imitating Si Doping Profiles—'Simitation'

$L_G = 70$ nm Ge pFETs with good short channel behavior require careful optimization of the source/drain doping profiles and the halo implants. A good starting point for this engineering exercise can be found by imitating the doping profiles in an optimized $L_G = 65$ nm pFET silicon technology. The calibrated Monte Carlo ion implant simulator is obviously a very useful tool to obtain these Ge I/I conditions. Therefore, combining 1-dimensional SIMS analysis with the MC simulator in the following paragraphs, we will try to match the $L_G = 65$ nm Si doping profiles with

Ge conditions. Note that this approach will only provide a starting point, not a complete set of optimized ion implant conditions for Ge for two reasons. Firstly, matching the chemical dopant concentration observed in Si with as-implanted dopant profiles in Ge does not—in general—yield identical charge carrier profiles. However at the dopant concentrations generally found in scaled pFETs (below 1×10^{19} cm^{-3} for n-type dopants), diffusion of dopants in Ge is rather small, as discussed in Chap. 2. As a result, the as-implanted dopant profiles in Ge will be a good approximation of the final dopant distribution (after the activation anneal). Secondly, the specific material properties of Ge will most likely require different dopant distributions for optimal pFET performance. For example, Ge's smaller band gap may require a smaller electric field in specific regions to reduce junction leakage. Keeping these limitations in mind, the different I/I processing steps are discussed (detailed conditions are in Table 3.3):

- *Well doping*—The higher atomic mass of Ge results in broader as-implanted profiles, compared to Si, as mentioned before. As a result, when attempting to obtain the same deep well profile as in Si, the broader P profile results in a too high channel doping. To avoid this, the deep well dose was lowered. In combination with matching conditions for the APT (Anti Punch Through) and VTA (V_T Adjust) well implants, the resulting P profile in Ge is quite close to its counterpart in Si, considering that the actual device is fabricated in the top 100 nm of the wafer (Fig. 3.6(a)). While using As for the deep well implants would allow matching the Si deep well profile, its higher atomic mass would cause the top 500 nm Ge to be amorphized as a result of the I/I at the required conditions, which is obviously not desired.
- *Halo doping*—The Si halo reaches a maximum As concentration of about 1×10^{19} cm^{-3} at a depth of 20–30 nm. The P halo implants (30–40–60 keV) which were successfully used in the $L_G = 125$ nm Ge pFETs [105] are all broader than the Si target halo profile, while not reaching the same peak concentration. For this reason an 80 keV As halo was chosen, matching the Si target profile in the first 20–30 nm and reaching the same peak concentration. As the extension depth is about 20–30 nm for a 65 nm technology, the slightly deeper tail of this halo implant is expected to have limited or no influence on the short channel behavior, although junction leakage might increase.
- *Extension doping*—While in the Si pFET, the as-implanted B extension profile is very shallow (the B concentration drops below 1×10^{19} cm^{-3} at a depth of only 12 nm), the 1050 °C spike anneal causes significant in-diffusion. Consequently, the as-implanted profile is not relevant for this exercise. In Ge, B is largely diffusionless up to 700 °C (see Sect. 2.2.2). Considering even lower 550 °C Ge junction anneal, the as-implanted B profiles in Ge can be considered a good approximation of the final B distribution. As shown in Fig. 3.6(c), a 2 keV, 8×10^{14} cm^{-2} B implant into Ge yields the same junction depth as its Si equivalent (measured using SIMS to account for the aforementioned B diffusion during the spike anneal). Note that the Ge extensions have a higher B concentration near the surface, which will give rise to a lower pFET series resistance assuming this extra B is electrically active.

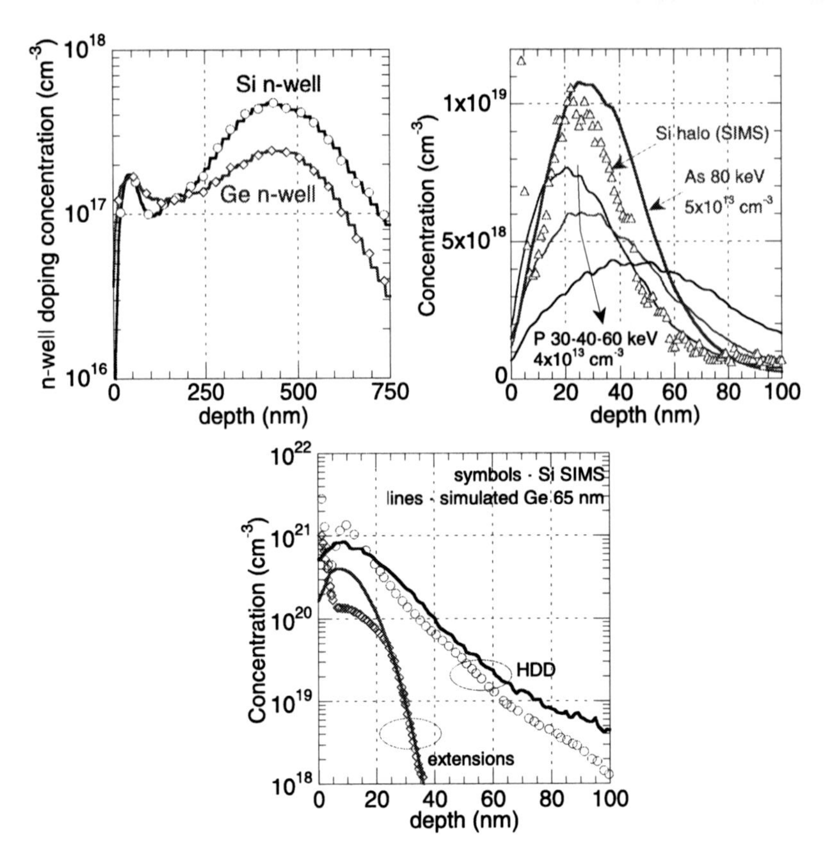

Fig. 3.6 Simulated or measured ion implant conditions for a Si $L_G = 65$ nm pFET flow and the matched Ge implant conditions (see also Table 3.3)

- *HDD doping*—Finally a 6 keV, 3×10^{15} cm^{-2} B implant is shown to yield a similar profile as its Si HDD equivalent. However, a preceding Ge preamorphization step is required in order to improve the electrical B activation and prevent an excessive ion-channeling tail (this was included in the MC simulation in Fig. 3.6(c).

3.4.2 Experimental Details

Ge pMOSFET devices were fabricated on 200 mm diameter, (100)-oriented Si wafers on which a relaxed Ge layer was grown to a thickness of 1.5 μm. The wafers were first annealed at 850 °C for 3 min to reduce the threading dislocation density to approx. 2×10^7 cm^{-2} [16]. The basic Si-compatible process flow is described in [33]. A phosphorus channel doping of about 3×10^{17} cm^{-3} was implanted, followed by deposited SiO_2 isolation. The Ge surface was passivated by a thin, partially oxidized epitaxial Si layer, as described in [16, 32], and capped with 4 nm HfO_2, after

Table 3.3 Ion implant conditions for the Si $L_G = 65$ nm pFET flow and matched Ge implant conditions

Process step	Ge 65 nm implants
n-well	P 540 keV, 1×10^{13} cm^{-2}, 7°
APT	P 190 keV, 1×10^{12} cm^{-2}, 7°
VTA	As 175 keV, 1×10^{12} cm^{-2}, 7°
Well anneal gate module	300 sec, 600 °C, N_2
Halo	As 80 keV, 5×10^{13} cm^{-2}, 15°
Extensions spacer module	B 2 keV, 8×10^{14} cm^{-2}, 0°
HDD	Ge 35 keV, 1×10^{15} cm^{-2}, 0°
	B 6 keV, 3×10^{15} cm^{-2}, 0°
Junction anneal	5 min RTA, 550 °C, N_2

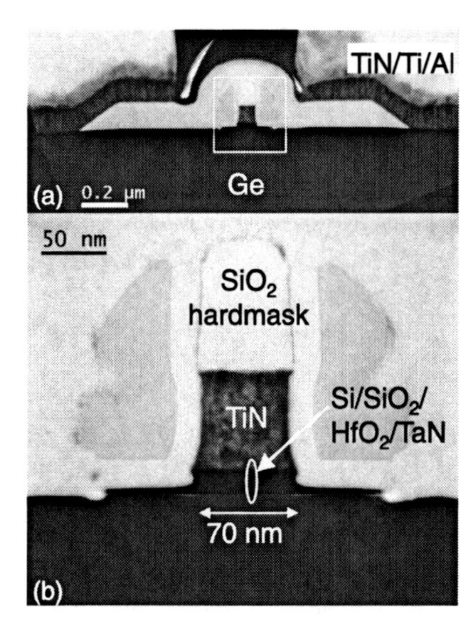

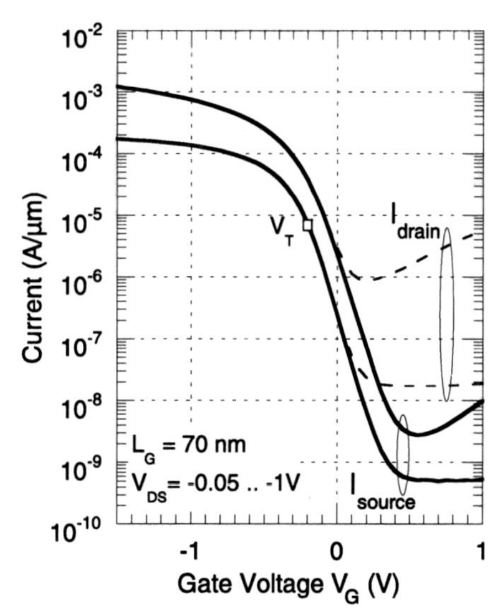

Fig. 3.7 Cross-sectional TEM image of a $L_G = 70$ nm Ge pMOSFET (**a**)–(**b**) and I_S–V_G/I_D–V_G characteristics for the 70 nm Ge pMOSFET with 'B only' extensions (**c**)

which a TaN/TiN metal gate is deposited. On the reference process, arsenic halos are implanted (80 keV, 5×10^{13} cm^{-2}, 15° tilt), followed by BF_2 extensions (11 keV, 8×10^{14} cm^{-2}). Spacer definition and HDD implants are followed by NiGe S/D formation (5 nm Ni deposited, 2-step RTP flow, [16]) and TiN/Ti/Al/TiN back-end processing [15].

Besides the reference flow just described (Process Of Reference—POR), devices were also fabricated with different implant conditions for the halo and extension doping. A 'shallow halo' was implanted using As, 60 keV, 4.5×10^{13} cm^{-2}, 15° tilt, a 'deep halo' was implanted using As, 100 keV, 5.5×10^{13} cm^{-2}, 15° tilt. The doses in all three conditions were changed slightly to maintain the same As peak

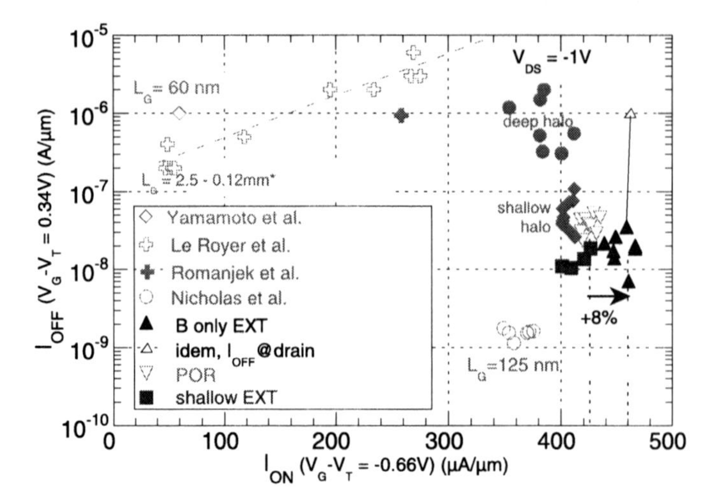

Fig. 3.8 I_{ON}–I_{OFF} relationship for the Ge pMOSFETs at $V_{DD} = 1$ V, evaluated at the source (various splits) and comparison with literature. A boost in I_{ON} is observed using 'B only' extensions

concentrations on the as-implanted profile. A 'shallow extension' was implanted by lowering the BF_2 implant energy to 9 keV while a 'B only' extension was fabricated using a B implant at 2.42 keV. The as-implanted profile B of the 'B only' condition is closely matching the 11 keV BF_2 POR extension I/I, while the 'shallow extension' condition using BF_2 matches a 2 keV B implant.

Note that the ion implant conditions used in the following sections are somewhat of a compromise between the existing Ge process flow, designed for L_G down to 125 nm, [105] and the 'Simitation' conditions discussed above. Still, essential elements such as the As halo, B extensions instead of BF_2 and a shallow extension split were included and will be discussed in detail.

Figure 3.7(a)–(b) shows a cross-sectional TEM image of the Ge pMOS with a gate length of 70 nm. The small voids next to the spacers were formed during the NiGe process module. While the exact reason is still under investigation, they may have an negative impact on the devices' electrical characterization.

3.4.3 Electrical Characterization

3.4.3.1 General Analysis

Figure 3.7(c) shows the I_S–V_G and I_D–V_G characteristics for the 70 nm Ge pMOSFET with 'B only' extensions for a V_{DS} of −50 mV and −1 V. A saturation drive current (I_{ON}) of 467 µA/µm is obtained for $V_G - V_{T,sat} = -0.66$ V and $V_{DS} = -1$ V, with an OFF-state current (I_{OFF}) of 2×10^{-8} A/µm at $V_G - V_{T,sat} = 0.34$ V, evaluated at the source. While I_{OFF} is conventionally measured at the drain to

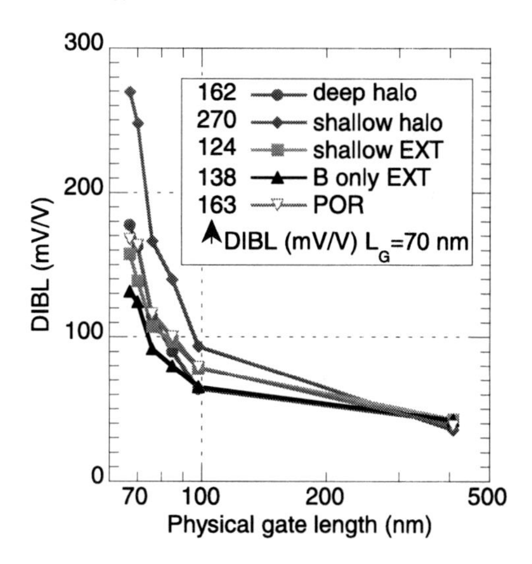

Fig. 3.9 DIBL vs. L_G for the Ge pMOSFET devices (various splits)

include drain-to-bulk diode leakage, Ge, with its smaller band gap suffers from higher diode leakage at the drain/well junction than equivalent Si devices [38] due to increased trap-assisted-tunneling mechanisms. Unfortunately the test devices discussed in this section have a rather large drain area of 148 μm^2, several orders of magnitude larger than those used for deep submicrometer devices. Measurements on devices with different drain areas have shown that this drain-to-bulk junction leakage varies with area. As such, evaluating I_{OFF} at the drain would result in an overestimation for realistic device dimensions. Therefore, I_{OFF} is measured at the source rather than estimating its value for small active areas. Nevertheless, drain-side I_{OFF} will be included in the further benchmarking of these Ge pFETs. The issue of drain-to-bulk leakage for Ge pMOSFETs will be discussed in more detail in Sect. 5.2.1.

The effective oxide thickness (*EOT*) of the gate stack is 1.4 nm. The gate leakage is less than 10^{-3} A/cm^2 (measured at $V_G = V_T - 0.6$ V). The V_T is about 0 V for long-channel devices and −200 mV for the 70 nm pFET (on target for high performance logic applications). This lower V_T is the result of the halo implants, which dominate for short devices (pFET V_T roll-up). Long channel V_T can be lowered by increasing the well doping, which will have limited or no impact on short channel devices where the halo doping dominates. Halo and well doping can thus be optimized together to obtain the desired flat V_T behavior as a function of gate length. Figure 3.8(a) shows the I_{ON}–I_{OFF} relationship for Ge pMOSFETs evaluated at the source. To compensate for the off-target V_T the V_G swing is shifted such that I_{ON} and I_{OFF} are measured at $V_G - V_{T,sat} = -0.66$ V and $V_G - V_{T,sat} = +0.34$ V respectively, as suggested by [19]. Available short channel Ge pFETs were added for comparison, including 60 nm devices with full NiGe source/drains [159] and devices fabricated on Germanium-On-Insulator substrates (GOI) [92, 121].

3.4.3.2 Extension Optimization

Comparing the different extensions implant conditions, two effects can be observed. First, reducing the BF_2 extension energy from 11 to 9 keV results in a decrease of I_{OFF} (Fig. 3.8) and *DIBL* from 163 to 124 mV/V (Fig. 3.9). This improved short channel behavior is of course linked to the smaller implant energy, which results in shallower extensions. In addition, the 'B only' extensions result in an 8 % I_{ON} boost, as compared to their BF_2 counterparts. At the same time a reduction in short channel effects is also observed (*DIBL* = 138 mV/V). Consequently, the I_{ON} boost cannot be caused by a reduced electrical gate length, as this would result in degraded short channel behavior. To pinpoint the cause, the devices' series resistance R_{EXT} was extracted using the method described in [140]. It was found that using the 'B only' extensions reduced R_{EXT} from 145 to 100 Ω µm. Using a linear approximation, such a decrease in R_{EXT} would explain a 5 % I_{ON} boost, which is quite close to the observed value.

Linking these pFET device results to the blanket boron shallow junction experiments performed in Sect. 2.2.2 confirms that the co-implantation of boron with fluor in Ge degrades junction properties. In the junction experiment, R_{sh} increased from 153 to 425 Ω/sq. Considering an extension length of 70 nm on either side of the gate, the total source and drain extension resistance would be 68 and 24 Ω µm for the POR and 'B only' extensions respectively. The difference of 43 Ω µm between these two accounts almost fully for the observed improvement in pFET series resistance R_{EXT}. As such, it is clear that the co-implantation of B with F, although necessary in Si technology to reduce B diffusion, degrades Ge pFET performance and should be avoided if possible to obtain higher I_{ON}.

3.4.3.3 Halo Optimization

Comparing the different halo implant conditions (POR, shallow halo, deep halo) a degradation is observed for both deviations from the POR condition in Fig. 3.8. Relative to the POR condition, the 'deep halo' condition significantly increases drain-to-bulk junction leakage, leading to 1–2 orders of magnitude higher I_{OFF}, while *DIBL* is almost unchanged for the $L_G = 70$ nm pFET. This clearly signifies that this halo implant condition results in a too deep halo, increasing the electric field and, as a result, the junction leakage while doing little to further improve short channel control. For the shallower halo, *DIBL* is severely degraded, leading to a mild increase in I_{OFF} in combination with a small I_{ON} penalty, indicating a too shallow halo placement.

3.4.4 Benchmarking

An important question for Ge is whether its high bulk mobility translates into high performing short-channel devices, as compared to Si. For aggressively scaled devices, carrier velocity saturation can be a key limiter for device performance. However, [109] has shown that hole velocity is higher in Ge compared to Si (maintaining similar parasitic effects). Here, we will compare the $L_G = 70$ nm Ge pFETs

Table 3.4 Comparison of key parameters for a $L_G = 70$ nm Ge pMOSFET and the ITRS specifications for the corresponding technology node (physical $L_G = 65$ nm) showing a 50 % higher I_{ON} for the Ge pFET

	ITRS 2002, [73] $L_G = 65$ nm	Ge pFET $L_G = 70$ nm
Physical gate length L_G (nm)	65	70
Equivalent oxide thickness EOT (nm)	1.3–1.6	1.4
Power supply voltage V_{DD} (V)	1.2	1.2
Saturation drive current I_{ON} (μA/μm)	400–450	622
saturation source off-state current I_{OFF} (A/μm)	2.4×10^{-8}	2.1×10^{-8}
Saturation drain[a] I_{OFF} (A/μm)		9×10^{-7}
Saturation I_{ON} at $V_{DD} = 0.95$ V (μA/μm)		432

[a]Overestimation for realistic device dimensions due to large drain area of 148 μm^2

with 'B only' extensions with strained Si, using the benchmarks proposed in [19]. Figure 3.10(a) shows the intrinsic gate delay as a function of the I_{ON}–I_{OFF} ratio for Ge pFETs (our devices, [105]) and a strained Si reference with similar EOT ($I_{ON} = 422$ μA/μm, [154]). The 70-nm Ge device offers a significant improvement over the 80-nm strained Si device for any I_{ON}–I_{OFF} ratio up to 4×10^4. A second metric is the intrinsic gate delay as a function of L_G (Fig. 3.10(b)). Also here, a certain benefit of Ge with respect to strained Si can be observed down to $L_G = 70$ nm. The third metric proposed in [19] is the linear subthreshold slope as a function of L_G. For our Ge devices, this quantity is rather constant at 120 mV/V. This high constant value indicates that further optimization of the gate stack is required before any conclusions can be drawn regarding the short channel control using this metric. The fourth benchmark (energy delay product as a function of L_G) is not discussed, as it will show the same as the second benchmark since we have benchmarked all devices at the same supply voltage and at similar EOT.

Comparing our Ge pMOS with the ITRS specifications for $L_G = 65$ nm (130 nm node, [73]) demonstrates the performance advantage. Table 3.4 summarizes the data and benchmarking conditions of this comparison. The Ge device exceeds the ITRS I_{ON} requirements by almost 50 %, maintaining a similar I_{OFF}, as measured at the source. A second comparison shows that the Ge pMOSFET can reach this ITRS specification ($I_{ON} = 432$ μA/μm) at a reduced V_{DD} of 0.95 V, yielding a 40 % reduction in active power dissipation, thanks to the V_{DD}-scaling ($P = fCV_{DD}^2$). This performance improvement is obtained despite the still slightly larger physical gate L_G for the Ge devices.

The significance of this benchmarking exercise is that is indicates that the higher mobility of Ge indeed translates into a higher drive current I_{ON}, even for sub-100 nm devices. Still, as more techniques arise to further strain Si devices, Ge FETs will, in turn, also require strain-boosters to keep outperforming (strained) Si: with a channel stress in excess of 1 GPa, the ballistic hole velocity in Ge devices would be higher than the limit in strained silicon based on simulations, [3]. A second is-

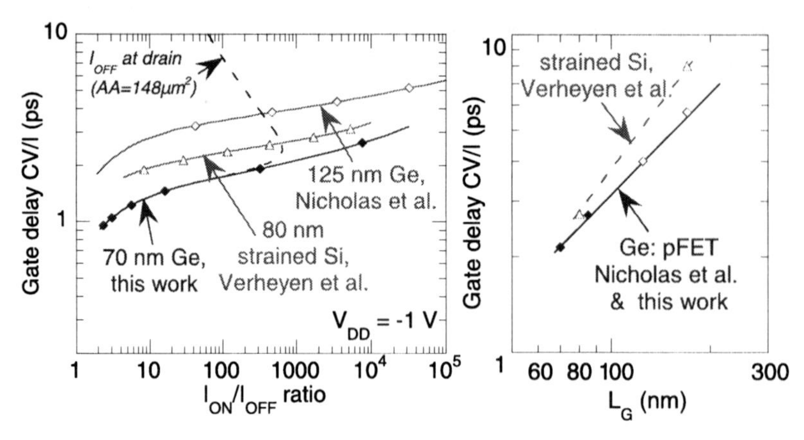

Fig. 3.10 Intrinsic gate delay as a function of I_{ON}–I_{OFF} ratio (**a**) and of gate length (**b**) for Ge (this work and [105]) and strained Si pMOSFETs [154]—Benchmarks from [19]

sue which needs to be investigated is the drain-to-bulk leakage in scaled devices. Addressing this issue, a scalability study of bulk Ge devices will be discussed in Sect. 5.2.1.

3.4.5 Conclusions

In this section, high performance Ge pMOSFETs were presented with physical gate lengths down to 70 nm. The ion implant process steps were designed to provide good short channel control in these devices by mimicking the doping profiles of an optimized 65 nm Si pFET technology using the calibrated Monte Carlo ion implant simulator. Different extension and halo implant conditions were compared. Firstly, the 80 keV As halo implant was shown to be optimal in terms of implant depth, with both shallower and deeper halo implants causing reduced short channel control and excessive junction leakage respectively. Secondly, reducing the extension depth and changing the implant species from BF_2 to atomic B resulted both in better short channel control and decreased pFET series resistance respectively. Benchmarking shows the potential of Ge to outperform (strained) Si as a pFET channel material well into the sub-100 L_G regime. The 70 nm devices outperform the ITRS requirements for I_{ON} for the corresponding node by about 50 %, maintaining a similar I_{OFF} measured at the source. In addition, the Ge device matched these ITRS specifications at a reduced V_{DD} of 0.95 V, resulting in a 40 % reduction in active power dissipation, owing to V_{DD} scaling.

3.5 Summary and Conclusions

In this chapter, a Monte Carlo simulator was calibrated to enable TCAD process simulations of ion implants into germanium substrates. Simulated as-implanted concentration profiles for B, P, Ga and As showed good agreement with measured data, obtained using Secondary-Ion Mass Spectroscopy measurements. Additionally, the thickness of the amorphous Ge layer created during the I/I as measured using Cross-sectional Transmission Electron Microscopy (TEM) or Spectroscopic Ellipsometry (SE), was shown to agree with the simulator's predictions, with typical deviations in the range of 0 to 10 %. Consequently, the calibrated MC simulator allows reliable simulations of as-implanted profiles and amorphization depths for common dopants in crystalline Ge.

In a second section, an analytical model was proposed based on the description of as-implanted dopant profiles (P, Ga, As) with dual Pearson distribution curves. This model allows a fast, straightforward way to obtain low-noise I/I profiles for these dopants in Ge and covers an energy-dose range comprising many conditions relevant for short channel Ge p- and nMOSFET development.

Then, using the calibrated Monte Carlo simulator, the ion implant steps required for a scaled $L_G = 70$ nm Ge pFET technology were designed, based on the optimized I/I profiles for a 65 nm Si pFET fabrication flow. Using many of those I/I conditions, high performance Ge pMOSFETs were fabricated with physical gate lengths down to 70 nm, allowing the comparison of different conditions for the extension and halo ion implants. An 80 keV (5×10^{13} cm^{-2}, 15° tilt) As halo implant was shown to be optimal in terms of implant depth for this technology. pFET series resistance was shown to decrease by changing the extension implant species from BF_2 to atomic boron, thus confirming the results obtained in blanket junction experiments (Chap. 2).

Finally, a benchmarking exercise showed the potential of germanium to outperform strained-silicon as a channel material well into the sub-100 L_G regime: 70 nm devices outperform the ITRS requirements for I_{ON} by about 50 %, maintaining a similar I_{OFF}, measured at the source. In addition, the Ge pFETs matched ITRS specifications with a 40 % reduction in active power dissipation, owing to V_{DD} scaling.

Appendix

A.1 Calibrated Parameters for TaurusMC

This appendix contains the calibrated parameters for the Taurus Monte Carlo Ion Implant simulator (using sprocess syntax, [128]).

```
PHYSICAL MODELS FOR GERMANIUM
pdbSetDouble ImplantData Germanium AtomicMass 72.61
pdbSetDouble ImplantData Germanium AtomicNumber 32
pdbSetDouble Germanium LatticeConstant 5.64613
pdbSetDouble Germanium LatticeDensity 4.41e22
pdbSetDouble Germanium AmorpGamma 1.0
pdbSetDouble Germanium AmorpDensity 1.1e22
pdbSetDouble Germanium AmorpThreshold 1.1e22
pdbSetDouble Germanium LatticeSpacing [expr pow(1/4.41e22,1.0/3.0)]
```

```
pdbSetString Germanium LatticeType Zincblende
pdbSetDouble Germanium MassDensity 5.35
pdbSetBoolean Germanium Amorphous 0
pdbSetString Germanium LatticeAtom COMPOSITION
pdbSetString Germanium Composition Component0 Name Germanium
pdbSetDouble Germanium Composition Component0 StWeight 1
pdbSetDouble Germanium CompoundNumber 1
pdbSetDouble Germanium DebyeTemperature 519
pdbSetBoolean Germanium ElectronicStoppingLocal 1
pdbSetDouble Germanium SurfaceDisorder 5e-4



NUMERICAL PARAMETERS FOR TaurusMC (please consult manual)
pdbSet MCImplant TrajectoryReplication 0
pdbSet MCImplant TrajectorySplitting 1
pdbSetDouble Germanium Phosphorus MaxSplits 8.0
pdbSetDouble Germanium Phosphorus MaxSplitsPerElement 1.0
pdbSetDouble Germanium Boron MaxSplits 8.0
pdbSetDouble Germanium Boron MaxSplitsPerElement 1.0
pdbSetDouble Germanium Arsenic MaxSplits 8.0
pdbSetDouble Germanium Arsenic MaxSplitsPerElement 1.0


Monte Carlo Implant paramerters implanted species (TaurusMC)
pdbSetDouble Germanium Phosphorus amor.par 1.0
pdbSetDouble Germanium Phosphorus casc.amo 1.0
pdbSetDouble Germanium Phosphorus disp.thr 15
pdbSetDouble Germanium Phosphorus casc.dis 15
pdbSetDouble Germanium Phosphorus surv.rat 0.75
pdbSetDouble Germanium Phosphorus casc.sur 0.75
pdbSetDouble Germanium Phosphorus MCVFactor 1.0
pdbSetDouble Germanium Phosphorus MCDFactor 1.0
pdbSetDouble Germanium Phosphorus MCIFactor 1.0

pdbSetDouble Germanium Boron amor.par 1.0
pdbSetDouble Germanium Boron casc.amo 1.0
pdbSetDouble Germanium Boron disp.thr 15
pdbSetDouble Germanium Boron casc.dis 15
pdbSetDouble Germanium Boron surv.rat 0.225
pdbSetDouble Germanium Boron casc.sur 0.225
pdbSetDouble Germanium Boron MCVFactor 1.0
pdbSetDouble Germanium Boron MCDFactor 1.0
pdbSetDouble Germanium Boron MCIFactor 1.0

pdbSetDouble Germanium Boron casc.sat 0.02
pdbSetDouble Germanium Boron sat.par 0.02
THESE PARAMETERS makes B only partially amorphizing in Germanium,
damage saturates when 2% of lattice atoms have been
displaced. This number is based on LIMITED SIMS data
and should be considered an estimate.
pdbSetDouble Germanium Arsenic amor.par 1.0
pdbSetDouble Germanium Arsenic casc.amo 1.0
```

```
pdbSetDouble Germanium Arsenic disp.thr 15
pdbSetDouble Germanium Arsenic casc.dis 15
pdbSetDouble Germanium Arsenic surv.rat 0.9
pdbSetDouble Germanium Arsenic casc.sur 0.9
pdbSetDouble Germanium Arsenic MCVFactor 1.0
pdbSetDouble Germanium Arsenic MCDFactor 1.0
pdbSetDouble Germanium Arsenic MCIFactor 1.0
```

A.2 Model Parameters: Ion Implants into Crystalline Ge

Table 3.5 Model parameters for the analytical ion implant modeling (see Sect. 3.3.2)

Species	p_0	p_1	p_2	p_3	p_4
P	-1.00×10^{-2}	8.58×10^{-1}	-1.00×10^{-1}	6.45	5.14×10^{-1}
Ga	-1.00×10^{-2}	4.23×10^{-1}	-1.00×10^{-1}	7.42	2.40×10^{-1}
As	-1.00×10^{-2}	4.03×10^{-1}	-1.00×10^{-1}	7.53	2.11×10^{-1}

Species	p_5	p_6	p_7	p_8	p_9
P	-6.72×10^{-4}	-4.01×10^{-1}	1.87×10^{-3}	-1.34×10^{-5}	1.01
Ga	-1.77×10^{-4}	-8.74×10^{-1}	3.12×10^{-3}	-5.74×10^{-6}	1.23
As	-1.25×10^{-4}	-7.44×10^{-1}	2.95×10^{-3}	-6.40×10^{-6}	1.27

Species	p_{10}	p_{11}	p_{12}	p_{13}	p_{14}
P	1.79×10^{14}	4.90	-2.17×10^{-2}	7.94×10^{-5}	2.71
Ga	9.04×10^{13}	7.78	-2.03×10^{-2}	2.62×10^{-5}	2.52
As	8.99×10^{13}	7.01	-1.84×10^{-2}	2.79×10^{-5}	2.64

Species	p_{15}	p_{16}	p_{17}	p_{18}
P	4.23	3.06×10^{-1}	4.29×10^{-2}	4.58×10^{12}
Ga	5.47	3.25×10^{-1}	-1.00×10^{-4}	2.46×10^{12}
As	5.70	3.17×10^{-1}	-1.00×10^{-4}	2.29×10^{12}

Chapter 4
Electrical TCAD Simulations and Modeling in Germanium

In this chapter, a TCAD device simulator is extended to allow electrical simulations of scaled Ge MOSFETs.

4.1 Introduction

Technology Computer-Aided Design (TCAD) software is widely used to optimize and predict the electrical behavior of many semiconductor devices. However, for non-Si materials, only a limited set of well-calibrated physical models is readily available. Specifically for germanium, several models that are indispensable for sub-100 nm Ge MOSFET simulations are lacking. A consistent set of physical models would allow for reliable device simulations. Used in tandem with the calibrated ion implant simulator presented in the previous chapter, a parallel virtual processing line can be constructed. Complementing the experimental efforts, such a TCAD combo would allow to optimize and predict the performance of scaled germanium based MOSFETs.

With this in mind, the first goal of this chapter will be to extend a commercial TCAD device simulator to allow for electrical simulations of sub-100 nm germanium pMOSFETs. Specifically, models for impurity scattering, mobility reduction in high electric fields, Shockley-Read-Hall recombination, trap-assisted tunneling and band-to-band tunneling will be implemented, based on experimental data and published results. As a second step, these models will be used to create a virtual processing line (combining process- and electrical simulations) for Ge pMOSFETs with L_G down to 70 nm, enabling detailed comparison with electrical measurements on actual devices.

In Ge technology, the electrical passivation of the interface between the semiconductor and the gate dielectric has proven to be a key challenge, as the density of interface states (D_{IT}) is often several orders of magnitude higher than that of the industry-standard Si/SiO_2 interface [7, 96, 141, 142]). Various measurement techniques were successfully extended to allow for a reliable extraction of the interface trap spectrum in Germanium FETs (e.g. the full conductance method [95]).

G. Hellings, K. De Meyer, *High Mobility and Quantum Well Transistors*,
Springer Series in Advanced Microelectronics 42, DOI 10.1007/978-94-007-6340-1_4,

Obviously, the performance of MOSFET transistors implementing germanium as a channel material can be severely degraded because of these interface traps [149]. As such, a technique is needed to quickly estimate the degradation arising from a given interface trap spectrum (D_{IT} profile as a function of energy, relative to the band edges). This would render a more direct link between the D_{IT} (in particular in certain regions close to the band edges) and corresponding MOSFET performance.

Therefore, the third goal of this chapter is to develop a methodology that allows to quickly study the electrostatic degradation, given a certain D_{IT} spectrum, of key performance parameters such as the subthreshold slope SS, drive current I_{ON}, and OFF-state current I_{OFF}. The relationship between interface traps and performance degradation is investigated for Ge n- and pMOSFETs and checked versus experimental results.

4.2 TCAD Models for a Germanium pMOSFET Technology

In this section, a commercial TCAD device simulator will be extended to allow electrical simulations of sub-100 nm germanium pMOSFETs. Mobility models will include impurity scattering and mobility reduction at high lateral and vertical electric fields. Models for generation-recombination mechanisms will be provided, covering Shockley-Read-Hall, trap-assisted tunneling and band-to-band tunneling. Finally, the interface traps are also included. Since the TCAD models used can be implemented in any device simulator, the actual model parameters (specific for Sentaurus sdevice, [127]) are placed in an appendix to this chapter. Also, note that some basic models (bandgap, density of states, dielectric constant, etc.) are not mentioned in this section, as their parameters are readily available in the literature and most TCAD software.

4.2.1 Modeling Mobility

A first set of models is related to carrier mobility. Germanium has a higher bulk mobility than silicon, a fact which is undoubtedly the main driver behind the renewed interest in Ge as a channel material for high-performance logic applications. However, when considering the carrier transport in a scaled Ge MOSFET, this argument becomes inevitably more complex because of the high doping levels and the high transversal and lateral electric fields. In this paragraph, we attempt to present a consistent set of models that should allow performing TCAD simulations of sub-100 nm Ge pMOSFET devices.

4.2.1.1 Phonon Scattering

This basic mobility model accounts only for phonon scattering. Since the higher Ge bulk mobility for both electrons (3900 versus 1400 cm^2/Vs for Si) and holes (1900

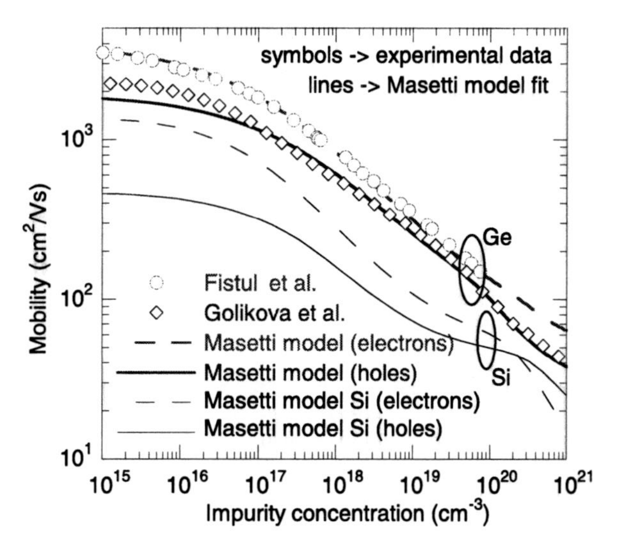

Fig. 4.1 Measured bulk mobility for electrons and holes in germanium [44, 50] and the TCAD fit, using the Masetti model. The Si mobility model is plotted for reference

versus 470 cm^2/Vs for Si) is already included in the default parameter set, it is not discussed further here.

4.2.1.2 Impurity Scattering

In doped semiconductors, the scattering of carriers by impurity ions results in a degradation of the carrier mobility. The model used in this work for carrier mobility in the presence of impurities was proposed by Masetti et al. [97] and is suited for semiconductors with an indirect bandgap such as Si and Ge. Note that this model is essentially an empirical fit to experimental data.

$$\mu_{dop} = \mu_{min1} e^{-\frac{P_c}{N_A+N_D}} + \frac{\mu_{const} - \mu_{min2}}{1 + (\frac{N_A+N_D}{C_r})^{\alpha}} - \frac{\mu_1}{1 + (\frac{C_s}{N_A+N_D})^{\beta}} \tag{4.1}$$

In this equation, which is the generalized implementation used in Sentaurus Device, N_A and N_D are the acceptor and donor concentrations, μ_{const} is the bulk mobility, the other parameters are used for numerical fitting (details in [127]).

In germanium, the doping-dependent mobility was measured for electrons and holes by Fistul et al. [44] and Golikova et al. [50], respectively. The parameters of the Masetti model were modified from their Si defaults to achieve an empirical fit with these experimental Ge data. The resulting mobility fit as a function of impurity concentration is plotted in Fig. 4.1. Note that the mobility in Ge remains higher than that in Si for any given impurity concentration.

4.2.1.3 Velocity Saturation

For high lateral electric fields, the carrier drift velocity is not proportional any more to the electric field. Instead, the carrier velocity saturates at a material and carrier

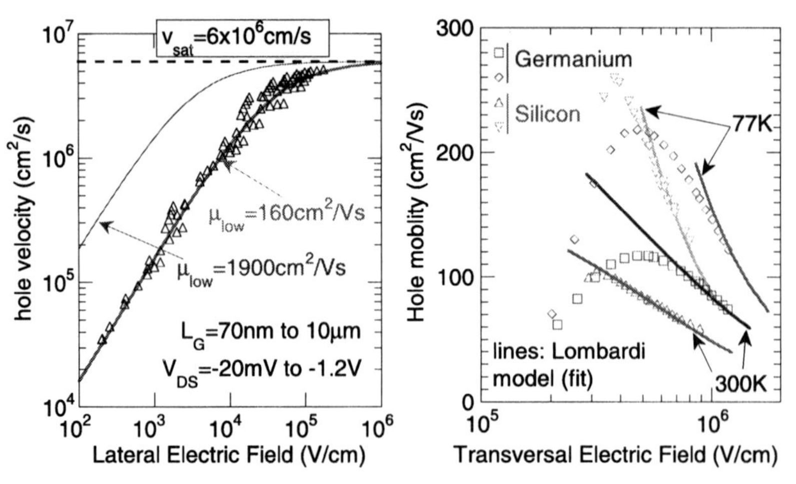

Fig. 4.2 (*left*) Measured average hole velocity in Ge pMOSFETs with L_G ranging from 70 nm to 10 μm as a function of average lateral electric field in the channel and the theoretical fit for this relationship using the Caughey-Thomas expression ([17] with $\beta = 1$ and $\mu_{low} = 160$ and 1900 cm^2/Vs). (*right*) Extracted hole inversion layer mobility as a function of transversal electric field at 77 K and 300 K for Si and Ge respectively, together with the TCAD fit using the enhanced Lombardi model [90, 127]

dependent velocity v_{sat}. Various sources state a numerical value of v_{sat} around 6×10^6 cm/s for both electrons and holes in Ge [85, 122, 138]. As verification, the average hole velocity in the inversion layer was extracted from I–V measurements on $L_G = 65$ nm Ge pMOSFETs using the method described in [147] (L_G ranging from 70 nm to 10 μm, $V_{DS} = -20$ mV to -1.2 V) and found to be in good agreement with the Caughey-Thomas expression [17]:

$$v(E) = \frac{\mu_{low} E}{[1 + (\mu_{low} E / v_{sat})^{\beta}]^{1/\beta}} \tag{4.2}$$

where μ_{low} is the low field mobility, E is the lateral electric field and β is a fitting parameter. Good agreement with our measurements was found for $\mu_{low} =$ 160 cm^2/Vs (Fig. 4.2). This value is significantly lower than the Ge bulk hole mobility (1900 cm^2/Vs) for a well doping concentration of 3×10^{17} cm^{-3}. Presumably, this is linked to the quality of the test material and/or gate dielectric interface. As Ge technology develops, material quality can be expected to improve. The value for β was not changed from the one used for holes in silicon ($\beta = 1$).

4.2.1.4 Acoustic Phonon/Surface Roughness Scattering

In the inversion layer of a MOSFET, the carriers are pushed against the semiconductor/dielectric interface by the large transversal (vertical) electric field. This additional source of scattering reduces the observed carrier mobility. It is dominated by acoustic phonon scattering at medium electric fields and by surface roughness scattering at high electric fields. Acoustic phonon scattering is very sensitive to temperature ($\sim T^{-1.75}$), while the scattering due to surface roughness is temperature-independent [139]. Thus, temperature-dependent measurements allow separating these two scattering mechanisms. Consequently, hole mobility was extracted at 77 and 300 K on Ge pFETs and on Si reference devices as a function of transversal electric field, using the split C-V method [100].

These measurements were then fitted using the Enhanced Lombardi Model, implemented in Sentaurus sdevice [127]. Note that this model is a TCAD implementation of the original model proposed by Lombardi et al. [90], allowing to implement both acoustic phonon and surface roughness scattering. The result of this exercise is given in Fig. 4.2. In the presence of high transversal electric fields, a satisfactory fit could be obtained by tuning the parameters of the Enhanced Lombardi Model.

The deviation at low electric fields is due to the fact that this model does not include Coulomb scattering. However, as will be shown in Sect. 4.3, including Coulomb scattering is not essential for the I_D–V_G and I_D–V_D simulations targeted in this work. Possibly, this is due to the fact that in the germanium pFETs analyzed in this work, the channel doping concentration already in the mid-10^{17} cm^{-3} range.

4.2.2 Modeling Generation-Recombination

The small bandgap of germanium (0.66 eV) is known to result in higher junction leakage [38, 39], compared to silicon devices. For this reason, an implementation of the various recombination mechanisms is indispensable for simulations of advanced Ge technologies.

4.2.2.1 Shockley-Read-Hall

The Shockley-Read-Hall (SRH) model describes recombination through deep-level defects in the bandgap. Typically, a description is used in terms of minority carrier lifetimes (τ_n and τ_p), dependent on the local defect concentration. Such a dependence arises from experimental data [151] and the theoretical conclusion that the solubility of a fundamental, acceptor-type defect is strongly correlated to the doping density [46, 127].

$$\tau_{SRH}(N_A + N_D) = \tau_{min} + \frac{\tau_{max} - \tau_{min}}{1 + (\frac{N_A + N_D}{N_{ref}})^{\gamma}} \tag{4.3}$$

In this equation, N_A and N_D are the acceptor and donor concentrations and τ_{max} is the maximum SRH lifetime. τ_{min}, γ and N_{ref} are used for numerical fitting to achieve a good agreement with published measurements of the minority carrier lifetime [47] and our own experimental leakage measurements on p+/n junctions [38].

It should be noted that the lifetimes reported in [47] are slightly higher than what would fit with our experimental data. This may be due to differences in sample morphology and remaining defects.

4.2.2.2 Trap Assisted Tunneling

In large electric fields, the lifetime of the minority carriers is reduced, through the mechanism of trap-assisted-tunneling (TAT). This is particularly relevant when simulating the abrupt junctions in scaled transistors. The model proposed by Hurkx et al. [69] to describe TAT was included in our simulations. It describes TAT in terms of a reduced field-dependent carrier lifetime τ_{TAT} as a function of the trap-assisted tunneling factor Γ_{TAT}:

$$\tau_{TAT} = \frac{\tau_{SRH}}{1 + \Gamma_{TAT}} \tag{4.4}$$

Note that the trap-assisted tunneling factor Γ_{TAT} is effectively reducing the minority carrier lifetime in the above formulation. Expressions for this parameter can be found in [127] or the original paper by Hurckx et al. [69]. In order to keep this text concise, they are not repeated here. Using this model, good agreement could be reached with leakage measurements on our p+/n junctions [38].

4.2.2.3 Band-to-band Tunneling

In even higher electric fields, the junction leakage is further increased through phonon-assisted band-to-band tunneling (BTBT). The model proposed by Schenk et al. [126] provides a simplified formalism for the purpose of device simulations. In this description, the generation/recombination rate is a function of the ratio of the local electric field in the structure and a critical electric field $F_c^{\pm}$ given by [127]:

$$F_c^{\pm} = B_{BTBT}(E_{g,eff} \pm \hbar\omega)^{3/2} \tag{4.5}$$

This expression contains the bandgap of the material and a pre-factor B_{BTBT} containing the effective mass for tunneling. For Ge, B_{BTBT} was scaled with respect to its Si default value, based on the ratio of the effective masses in both materials. As in the previous paragraphs, the lengthy expressions for B_{BTBT} and related parameters are not copied into this text.

4.2.2.4 Generation-Recombination—Models

To illustrate the contribution of the different leakage mechanisms (SRH, TAT and BTBT) as a function of the active doping concentration at the p+/n junction, ideal one-sided junctions were simulated. This results in 3 lines in Fig. 4.3, each considering one of the above-mentioned leakage mechanisms:

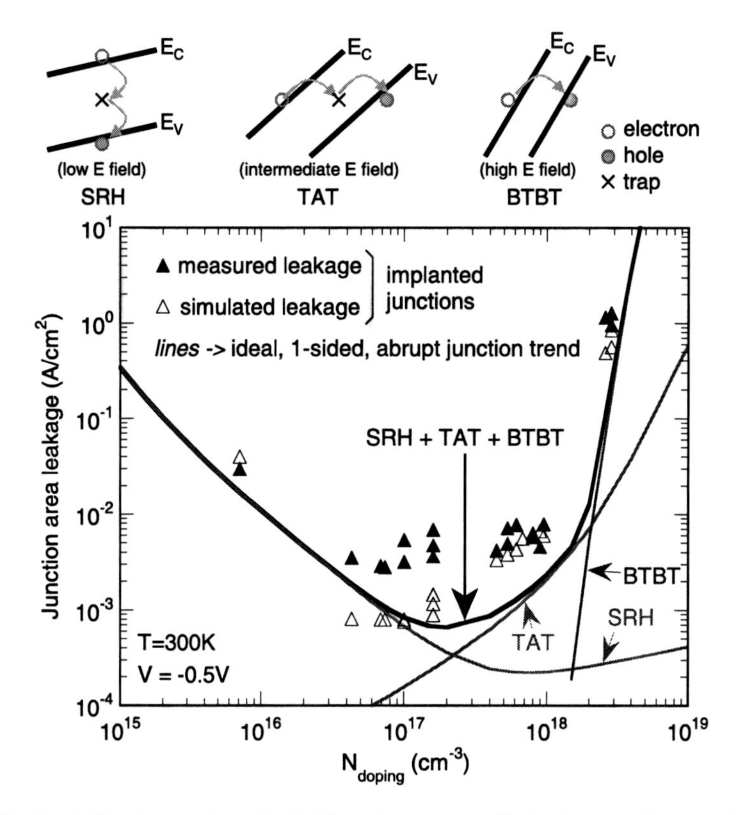

Fig. 4.3 (*lines*) Simulated leakage for 1-sided, abrupt, p+/n diodes in germanium as a function of n-type doping concentration, considering Shockley-Read-Hall (SRH), Trap Assisted Tunneling (TAT) and Band-to-band Tunneling (BTBT). Measured (*solid symbols*) and simulated leakage (*empty symbols*) for various implanted junctions in germanium. Doping and impurity profiles were simulated with TaurusMC in sprocess [128], preceding electrical simulations in sdevice [127]

- At low doping concentrations ($N_{doping} < 10^{17}$ cm^{-3}), the SRH mechanism is dominant. Electric fields in the space charge region remain rather small. In this regime, the junction area leakage decreases for increasing doping concentration due to the reduced volume of the space charge region.
- At intermediate doping concentrations ($2 \times 10^{17} < N_{doping} < 2 \times 10^{18}$ cm^{-3}), the TAT mechanism is dominant, as the electric field in the space charge region increases. The trap-assisted tunneling factor Γ_{TAT} (which depends rather strongly on the electric field) quickly increases. As a result, junction area leakage reaches a minimum at $N_{doping} = 2 \times 10^{17}$ cm^{-3}, after which it quickly increases.
- At high doping concentrations (2×10^{18} cm$^{-3} < N_{doping}$), higher doping concentrations enlarge the electric field in the space charge region even more. The BTBT mechanism kicks in, causing a sharp increase in junction leakage current density. Note that this mechanism is independent of local defect concentration.

With these considerations in mind, all three leakage mechanisms should typically be included when simulating sub-100 nm Ge pFETs. Specifically, TAT and BTBT are indispensable for the highly doped junctions found in these devices around the halos and around the extension regions. A correct estimation of I_{OFF} will depend on the inclusion of these models.

4.2.2.5 Generation-Recombination—Experimental

To check the applicability of the models discussed above in simulations of scaled Ge pMOS devices, the junction leakage was measured for various p+/n junctions [38]. For these fabricated junctions, the dopant profiles were also simulated based on the implant conditions using the calibrated Monte Carlo implant simulator, discussed in Chap. 3. The resulting dopant distributions were then fed to the device simulator, simulating the leakage current density. This approach allows simulating the leakage for these junctions, taking into account ion implant tails and the local impurity concentration.

In this manner, pairs of measured and simulated junction leakage are obtained. This comparison is shown in Fig. 4.3 with the solid and empty symbols showing measured and simulated data respectively. A good agreement is found between the model and the experimental data, although measured junction leakage is still somewhat above the simulations for N_{doping} around $10^{17}\mathrm{cm}^{-3}$. A possible explanation for this may be that the TAT leakage is enhanced in this region because of additional deep-level defects. However, further investigation is needed to confirm this mechanism.

Note that the TCAD models for recombination are largely based on an empirical description. As such, even though we found good agreement with experimental data on various p+/n junctions, caution may be advised.

4.2.3 Modeling Interface Traps

A detailed discussion on interface traps and their influence on device performance will be given in Sect. 4.4. However, since the Ge pMOSFETs discussed in the following section include a certain distribution of interface traps, they are quickly mentioned here for the sake of completeness.

The electrical passivation of the interface between the high-κ dielectric and the channel is a key challenge for Ge technology. A high density of active interface traps will severely harm the performance of any Ge MOSFET. Various attempts have been made to find a process scheme which results in such a passivation and promising results have been obtained with (among others) an ultrathin Si-layer [32, 100] and GeO_x [7, 23].

In the following section, we will focus on Ge pMOSFETs with an ultrathin Si-passivation layer. For this processing scheme, the density of interface traps as a function of the energy in the bandgap was extracted with the full conductance method [95]. The extracted profile of interface traps was included in our TCAD simulations: Fig. 4.7(a) shows the measured data for a deposited Si-layer of 8 monolayers (ML), and the continuous profile used in the device simulations.

4.3 Electrical TCAD Simulations—65 nm Ge pMOSFET Technology

In this section, the models discussed in Sect. 4.2 will be used in electrical simulations of Ge pMOS devices with L_G down to 65 nm. First, simulated drain, source and bulk currents as a function of V_G and V_D are compared with electrical measurements. Second, the effect of changes in halo and extension (LDD) energies on short-channel control, drain-to-bulk leakage and drive current is investigated. Finally, the reference device performance will be shown to increase if the interface traps are removed from the simulations.

4.3.1 Simulator Setup

Since such electrical simulations are quite sensitive to the dopant profiles of the simulated device, these were obtained with the calibrated Monte Carlo ion implant simulator discussed before in Chap. 3 and Sentaurus Process [128]. The result of these simulated ion implants, the other processing steps and the numerical meshing is shown in Fig. 4.4, together with a TEM-image of the 65 nm device. The process flow and experimental details for the 65 nm Ge pMOSFET technology can be found in Sect. 3.4.2 and [56]. A SIMS measurement of the HDD region of the device shows the simulated and experimental dopant concentrations for boron, the arsenic halo and the phosphorus well. The noise on the measured As profile is quite large, which is due to mass interference during the SIMS measurement. The phosphorus concentrations in our samples are quite close to the noise floor of the SIMS measurement, giving rise to a larger discrepancy for this species. Additionally, a surface concentration peak is observed for both the As and P profiles; however, this is an artefact of the SIMS measurement and can therefore be ignored. Note that ion channeling during the implant is taken into account by the simulator and is quite pronounced for the boron profile. The maximum electrically active concentration level for boron was set to 4×10^{20} cm^{-3}.

The electrical TCAD simulations were performed in sdevice [127]. Mobility models include impurity scattering, velocity saturation, acoustic phonon scattering and surface roughness scattering. Generation-recombination mechanisms are Shockley-Read-Hall, trap assisted tunneling and band-to-band tunneling. The aforementioned density of interface traps was included at the gate-to-channel interface. A Capacitance Equivalent Thickness (CET) of 1.73 nm, matching with experimental C-V measurements was also assumed.

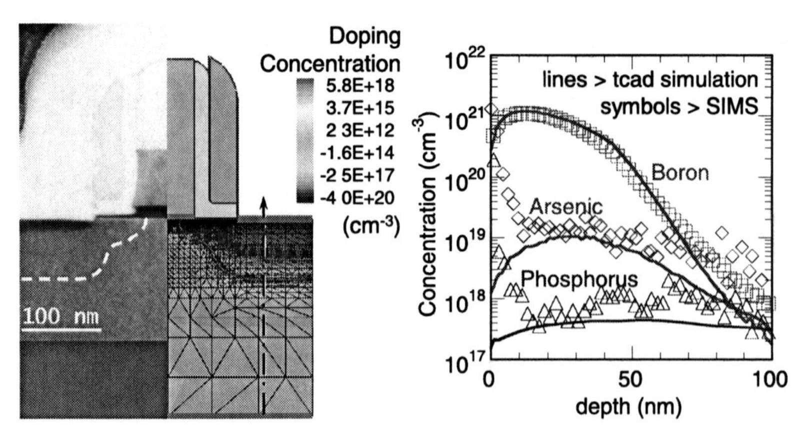

Fig. 4.4 (*left*) TEM image of the $L_G = 70$ nm Ge pMOSFET and the simulated structure, showing dopant distribution, contacts and meshing strategy used. (*right*) Measured (SIMS) and simulated chemical concentration profiles for boron, arsenic and phosphorus in the HDD region of this device

4.3.2 I_D–V_G and I_D–V_D Simulations

In Fig. 4.5(a), (c), (e), (f) the simulated transistor terminal currents are plotted together with those measured on a representative device as a function of V_G and V_{DS} for $V_{DS} = -20$ mV and -1 V and $L_G = 70$, 100, 250 and 1000 nm respectively. The simulated I_{DS}–V_{DS} current is plotted in Fig. 4.5(b), (d) together with the measured one for $V_{GS} - V_T$ ranging from 0 to -1.2 V for the $L_G = 70$ and 100 nm devices. The simulated curves show a very good agreement with the measured ones: e.g. for the 70 nm device at $V_{GS} - V_T = V_{DS} = -1$ V a drain current of 830 μA/μm is observed (simulated: 860 μA/μm), while an experimental drain-induced-barrier-lowering (DIBL) value of 140 mV/V is comparable to the simulated 130 mV/V. The drain-to-bulk junction leakage is also reproduced (e.g. 2.24×10^{-6} A/μm (measured) vs. 2.79×10^{-6} A/μm (simulated) for $L_G = 70$ nm).

The TCAD models capturing mobility at high lateral and transversal electric field are indispensable in obtaining a good agreement between the simulated and measured drive current and are observed to be sufficient for L_G down to 70 nm. In even smaller devices, the phenomenon of velocity overshoot, which is predicted by Monte Carlo simulations [35] may have to be included as well. In turn, the generation-recombination models (especially the TAT and BTBT models) are needed to capture the high drain-to-bulk leakage and Gate-Induced-Drain-Leakage (GIDL).

4.3.3 Alternative Implant Conditions

Since TCAD software is often used to optimize and predict the electrical effect of any change in the processing scheme, the validity of such an extrapolation has to be closely monitored. Here, an example case is presented, evaluating the effect of two

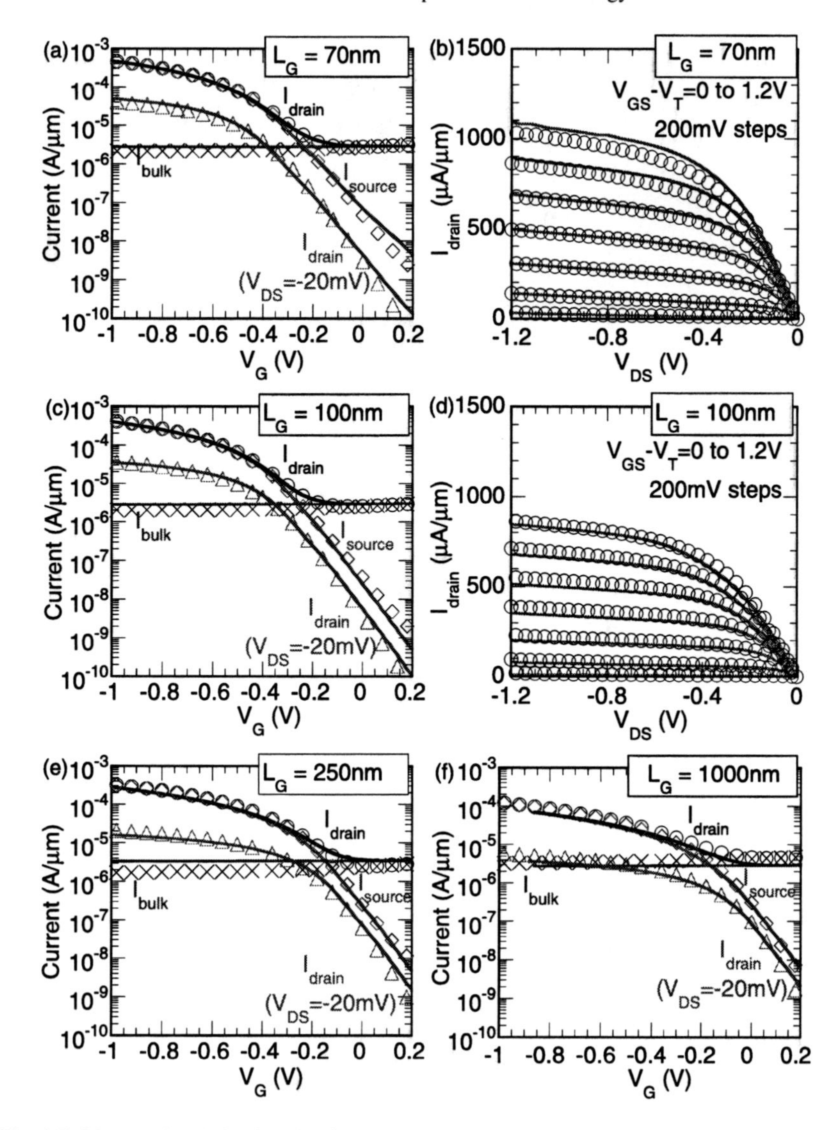

Fig. 4.5 Measured and simulated drain, source and substrate currents as a function of V_G for Ge pMOS devices with $L_G = 70$ to 1 µm at $V_{DS} = -20$ mV and -1 V or as a function of V_D for devices with $L_G = 70$ and 100 nm for $V_{GS} - V_T = 0$ to -1.2 V. In all plots, *lines* are simulated data while the *symbols* correspond to measurements

perturbations on the standard germanium $L_G = 65$ nm transistor flow: (1) increasing the extension implant energy from 2 to 2.4 keV and (2) increasing the implanted halo dose from 5×10^{13} to 6.5×10^{13} cm^{-2}.

For each of the perturbations, I_{ON}, V_T, *DIBL* and drain-to-bulk junction leakage were extracted from the measurements and simulation. This is illustrated in Fig. 4.6.

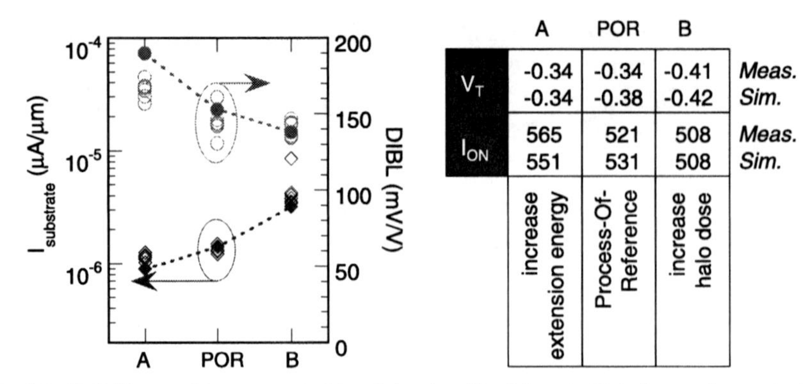

	A	POR	B	
V_T	-0.34	-0.34	-0.41	*Meas.*
	-0.34	-0.38	-0.42	*Sim.*
I_{ON}	565	521	508	*Meas.*
	551	531	508	*Sim.*
	increase extension energy	Process-Of-Reference	increase halo dose	

Fig. 4.6 (*left*) Measured (*empty symbols*) and simulated (*solid symbols*) drain-to-substrate leakage and DIBL for the standard Ge pMOS device with $L_G = 100$ nm and the effect of two process variations: (*A*) increasing extension implant energy from 2 keV to 2.4 keV; (*B*) increasing halo implant dose from 5×10^{13} to 6.5×10^{13} cm^{-2}. (*right*) Effect of these process variations on V_T and I_{ON} ($V_{GS} - V_T = -0.66$ V)

Comparing the measured quantities with the simulated ones shows that the electrical effect of these processing changes can be accurately predicted by the simulator.

- (A) Increasing the extension implant energy, results in a deeper, less abrupt junction and reduced short channel control. As a result, more DIBL is observed while drain-to-bulk leakage is reduced somewhat in the experimental results. These effects are also captured in the simulations: the higher *DIBL* and I_{ON} and lower drain-to-bulk leakage are reproduced. The predicted small change in V_T is not observed in the measurements.
- (B) Increasing the implanted halo dose is expected to improve short channel control. Indeed, a lower *DIBL* value is observed for this experiment, although other parameters are also affected: drain-to-bulk leakage increases sharply, I_{ON} is degraded and V_T shifts to a more negative value. Also for this perturbation, the observed effects are reproduced by the simulations.

4.3.4 Interface Traps

In this third part, the effect of interface traps is illustrated. In the previous simulations, the interface trap spectrum as a function of energy (Density of Interface Traps—D_{IT}) was included, as measured with the full-conductance method. Of course, the D_{IT}-spectrum is specific to the passivation process available at this moment. This raises the question what the characteristics would be of a hypothetical Ge pMOSFET without interface traps. For this reason, the 100 nm gate-length Ge pMOS was simulated also without these interface traps, keeping all other models unchanged. As can be seen in Fig. 4.7, the interface traps have a visible effect on the transistor's switching characteristics. Notably, removing traps reduces *DIBL* from 90 to 58 mV/V and reduces the subthreshold slope from 120 to 80 mV/dec. As

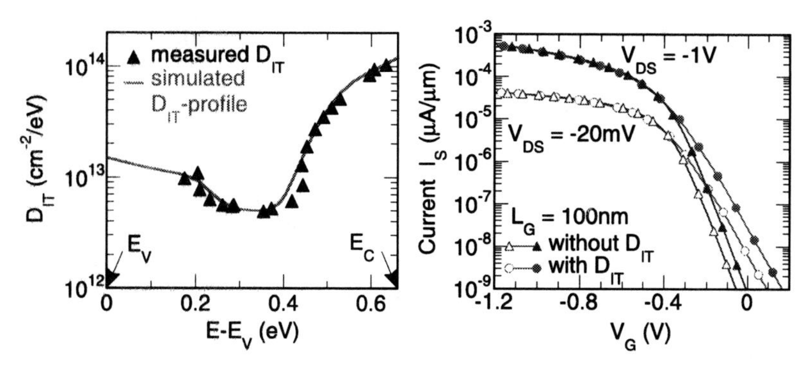

Fig. 4.7 (*left*) Measured interface trap density for the passivation recipe used in the $L_G = 65$ nm Ge pMOSFET technology as a function of energy, as extracted with the full-conductance method [95]. (*right*) Simulated source current as a function of V_G with and without these interface traps, showing improved short-channel control in the device without interface traps

such, further reduction of the interface trap density will be absolutely necessary to obtain a better electrostatic gate control. Additionally, the dependency of the carrier mobility on the transversal electric field can be expected to be different. As such, I_{ON} can be expected to increase, for an improved interface passivation quality.

4.3.5 Conclusions

In this section, a commercial TCAD device simulator was extended by adding a consistent parameter set for simulations of scaled Ge pMOSFETs. The model parameters for generation-recombination mechanisms (Shockley-Read-Hall, Trap-Assisted-Tunneling and Band-to-Band tunneling), mobility models (impurity scattering, mobility reduction at high lateral and transversal field) were adapted based on available experimental data. Electrical simulations of Ge pMOS devices using these models and parameters were found to be in good agreement with measured I–V curves for various bias conditions and gate lengths ranging from 70 nm to 1 μm. Finally, the electrical effect of changes in halo dose and extension implant energy was simulated and shown to be in good agreement with the experimental results.

4.4 Impact of Interface Traps MOS Performance

In the following section, a methodology is presented which allows to quickly study the electrostatic degradation, given a certain profile of interface traps (D_{IT}), on key MOSFET parameters such as its subthreshold slope (*SS*), drive current (I_{ON}) and off-state current (I_{OFF}). The relationship between interface traps and performance degradation is investigated for n- and p-channel MOSFETs, with respect to the gate length and the dielectric thickness, and is checked versus experimental results on Ge pMOSFETs.

4.4.1 TCAD Modeling and Electrical Characterization

The effect of interface traps on germanium n- and pMOSFETs is investigated using TCAD simulations (Sentaurus Device, [127]). The mobility and generation-recombination models were calibrated in the previous sections and include the Ge bulk mobility, impurity scattering and velocity saturation at high lateral electric field. Generation-recombination models include the Shockley-Read-Hall mechanism and Trap-Assisted-Tunneling. Band-to-band Tunneling is not included in the simulations because this leakage component is typically independent of interface traps: as it dominates the drain-to-bulk leakage, it may hide part of the effect of D_{IT}.

Unless otherwise mentioned, simulations in this section were performed on n- and pMOSFETs with a gate length L_G of 250 nm, an equivalent oxide thickness (EOT) of 1 nm and a substrate doping level of 3×10^{17} cm^{-3}. The simulations use an internal discretization step of 10 meV for interface trap profiles as a function of energy. Note that these simulations only include the electrostatic effect of interface traps on the transistor characteristics (i.e. only the effect of charging/de-charging the traps). Based on [142], additional scattering at interface traps may reduce the carrier mobility. However, the inversion layer mobility depends also on other parameters than the density of interface traps and is linked to the specific properties of the semiconductor/dielectric interface [100]. In order to provide a generalized formalism, those effects are not included in this work. However, the presented methodology does allow investigating the relationship between measured carrier mobility and interface traps, as will be shown in Sect. 4.4.4. Note also that the interface trap profiles in this work are considered to be spatially uniform.

The Ge MOSFETs in this section (simulated and experimental) are characterized at a supply voltage of $V_{DD} = 1$ V. I_{ON} and I_{OFF} are extracted at a fixed offset voltage from the threshold voltage V_T ($-0.66 \times V_{DD}$ and $0.34 \times V_{DD}$ respectively), after [19]. The subthreshold slope SS is taken as the average slope between two points on the sub-V_T I_D–V_G curve (I_D at $V_T + 50$ mV and $V_T + 340$ mV for pFETs). For a correct estimation of I_{OFF}, drain junction leakage must be included [56]. However, this component is independent of interface traps, as mentioned before. Therefore I_{OFF} is measured at the source, where it is dominated by subthreshold leakage, thereby visualizing the effect of interface traps. Note that following the current definitions, I_{ON}, I_{OFF} and SS are also not affected by V_T shifts.

4.4.2 Uniform Trap Spectra

In the simplest case, the interface traps are considered uniformly distributed as a function of energy. This situation is represented schematically in Fig. 4.8(a), consisting of a constant concentration of donors and acceptors (resp. below and above mid-gap level). Note that the type of interface traps (donor, acceptor, double acceptor) does not influence the I–V characteristics of the MOSFET, except for the position of the threshold voltage V_T (i.e. changing from a donor to an acceptor trap is equivalent to adding a fixed negative charge which results only in a V_T-shift).

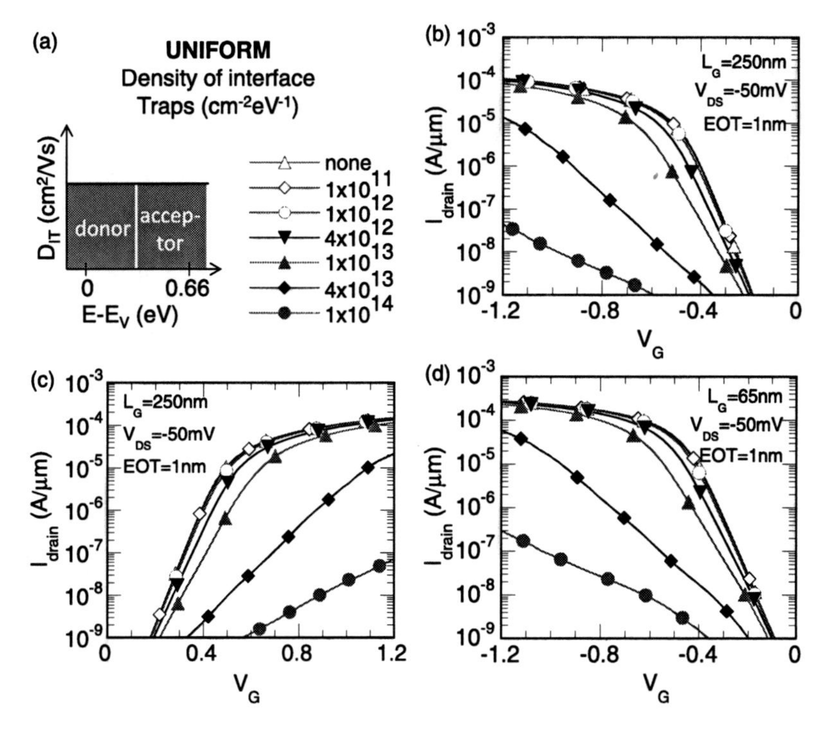

Fig. 4.8 (**a**) Schematic of energetically uniformly distributed interface traps. (**b**)–(**d**) Electrostatic effect of such trap profiles (varying trap density) on I_D–V_G characteristics of a $L_G = 65$ nm Ge pMOS, a $L_G = 250$ nm Ge nMOS and a $L_G = 250$ nm Ge pMOS

As such, the specific trap type has no influence on the V_T-corrected performance parameters used (*SS*, I_{ON} and I_{OFF}).

Using the TCAD simulator, the effect of such traps on the I–V characteristics can be investigated. Figure 4.8(b)–(d) shows simulated I_D–V_G curves for a Ge pMOS ($L_G = 65$ and 250 nm) and a Ge nMOS ($L_G = 250$ nm) for increasing D_{IT}. Clearly, uniform trap densities below 10^{12} cm^{-2} eV^{-1} do not result in any noticeable degradation of the I–V curves. On the other hand, uniform trap densities in excess of 10^{13} cm^{-2} eV^{-1} increasingly harm the switching properties of these transistors. Note also that the degradation is similar for the $L_G = 65$ and 250 nm devices (keeping all other parameters constant): i.e. a uniform trap density of 10^{13} cm^{-2} eV^{-1} increases the *SS* with about the same percentage, independent of L_G.

4.4.3 Non-uniform Trap Spectra

In actual devices however, the interface traps are not uniformly distributed as a function of energy as assumed in the previous section. Therefore, the effect of interface traps on key device performance metrics such as *SS*, I_{OFF} and I_{ON} is investigated,

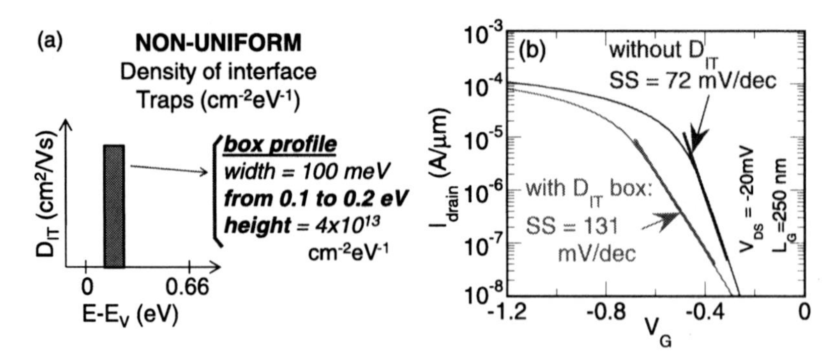

Fig. 4.9 (**a**) Schematic of energetically non-uniformly distributed interface traps (i.e. a 100 meV wide box-like D_{IT} profile defined by two parameters: peak concentration and position in the bandgap). (**b**) Electrostatic effect of this box-like D_{IT} profile on the I_D–V_G characteristics of a 250 nm Ge pMOS, yielding a 82 % increase in the linear subthreshold slope

depending on their position with respect to the valence band edge. A methodology is introduced to assess the degradation due to D_{IT}, focussing on the pMOS subthreshold slope. It is then extended to the other performance metrics and to nMOS.

4.4.3.1 Introduction

In Fig. 4.9, I_D–V_G curves were simulated for a Ge pMOS with a specific D_{IT}: a box-like trap profile with a height of 4×10^{13} cm^{-2}eV^{-1} between $E_V + 0.1$ eV and $E_V + 0.2$ eV. This specific trap profile results in a 82 % increase in linear subthreshold slope from 72 to 131 mV/dec, compared to the pMOS without any interface traps. Performing similar simulations with box-like trap profiles, the relative degradation of the subthreshold slope can be investigated as a function of the box height (trap concentration) and the box position (trap energy).

The results of this study are presented in a contour plot Fig. 4.10(a), (d). While the magnitude of the degradation indeed increases with the peak height, it is clear that the *SS* in Ge pMOS devices is degraded most by interface traps placed about 0.1 eV above the valence band edge (e.g. a box-like trap profile with a height of 10^{13} cm^{-2}eV^{-1}, placed at $E_V + 0.1$ eV results in a 20 % degradation in *SS*, while the same trap profile, placed 0.3 eV above E_V only results in a 5% *SS* degradation). Similar plots were also made for I_{OFF} and I_{ON} (Fig. 4.10(b), (c), (e), (f)). These show that the I_{OFF} of a pMOS can increase dramatically because of traps located between E_V and $E_V + 0.2$ eV. On the other hand, I_{ON} is affected more by traps located inside the valence band. These conclusions can be understood by considering the position of the Fermi Level E_F with respect to E_V in the pMOS in "ON" and "OFF" state (e.g. traps located just above E_V will 'slow down' the movement of the Fermi level E_F when changing the gate voltage from V_T to $V_T + 0.34$ V, since then E_F needs to move through the traps). Note also that I_{ON} is only mod-

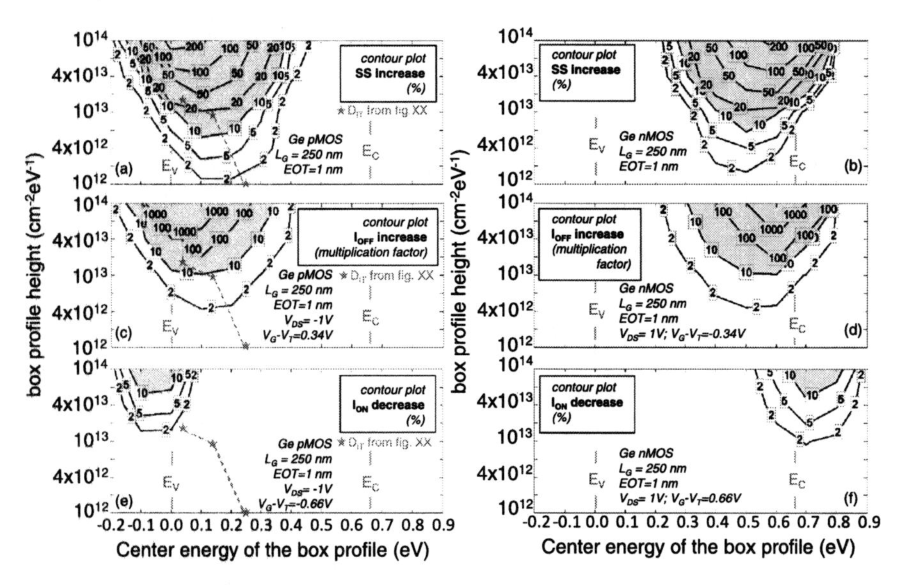

Fig. 4.10 Contour plot containing the relative degradation in the pMOS/nMOS subthreshold slope (**a**), (**d**), OFF-state current I_{OFF} (**b**), (**e**), and drive current I_{ON} (**c**), (**f**) due to a 100 meV wide box-like D_{IT} profile in function of its height (trap concentration) and position in the bandgap with respect to the valence band edge E_V. Simulations were performed for $L_G = 250$ nm Ge nMOS/pMOS devices with an *EOT* of 1 nm. Note the insensitivity to interface traps in the lower/upper part of the bandgap for nMOS/pMOS respectively

erately affected by rather high D_{IT} densities (e.g. a box profile with a height of 4×10^{13} cm^{-2} eV^{-1} placed at the worst position degrades I_{ON} by less than 10 %). This is due to the fact that the Fermi level E_F remains largely at the same position with respect to the valence band edge, once high inversion is reached. Note that, as mentioned before, only the electrostatic effect of interface traps was included here; other factors, such as the mobility should definitely be considered to fully assess the effect on I_{ON}.

For Ge nMOS, the same analysis was done, resulting in the contour plots in Fig. 4.10. As may be expected, these look very similar to their pMOS counterparts, albeit mirrored. As such, nMOS performance is mostly affected by traps located about 100 meV below the conduction band edge.

4.4.3.2 Methodology

These plots can now be used to assess the electrostatic degradation of Ge MOSFETs for arbitrary D_{IT} profiles. The methodology for this is depicted schematically in Fig. 4.11(a)–(d). A given D_{IT} profile can be approximated by adjacent box-shaped

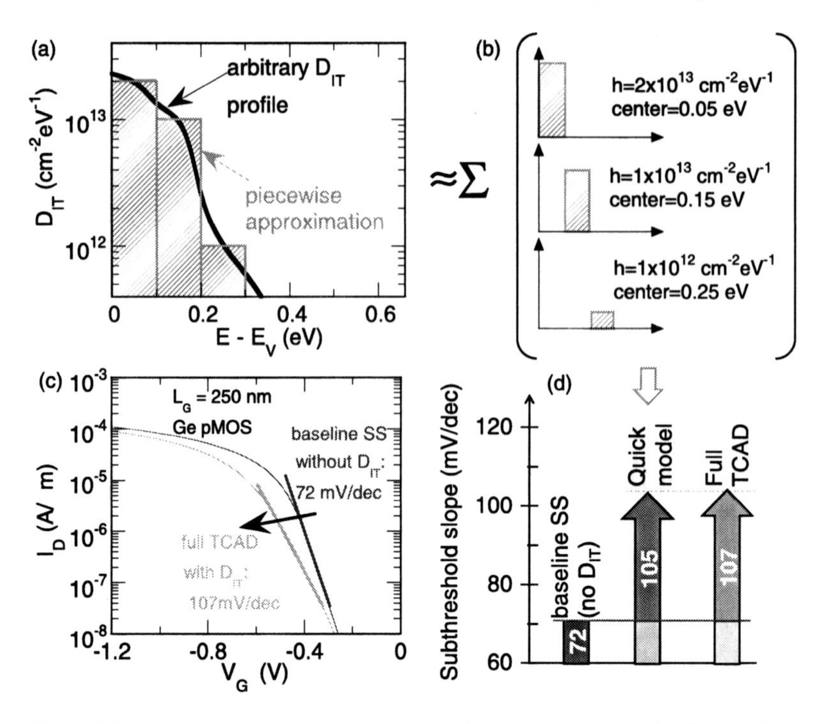

Fig. 4.11 (**a**) Piecewise approximation of an arbitrary D_{IT} spectrum using (**b**) three boxlike profiles The resulting relative degradation in pMOS SS I_{OFF} and I_{ON} are then calculated by summing the contributions from each of these boxes. The values used from Fig. 4.10 are also indicated there. (**c**) Simulated I_D–V_G curves with and without the D_{IT} profile from (**a**) on an $L_G = 250$ nm Ge pMOS device. (**d**) comparison between the model prediction (summing the three box-like profiles' contribution), and full TCAD device simulations including the original D_{IT} profile from (**a**)

trap profiles (Fig. 4.11(a)). For each of these, the *SS* degradation is given by the contour plots. The sum of the degradation of the different box-shaped trap profiles yields the total degradation. For the example in Fig. 4.11(a), the different boxes yield a total increase of $26\,\% + 16\,\% + 1\,\% = 43\,\%$ with respect to the baseline *SS* for the pMOS without any interface traps (or a predicted 105 mV/dec instead of 72 mV/dec for the pMOS with D_{IT}). Checking the validity of this method, the full D_{IT} profile from Fig. 4.11(a) was also simulated and yielded a *SS* of 107 mV/dec, a value very close to the model proposed in Fig. 4.11. Similarly, I_{OFF} is predicted to increase by a factor 40× (from 2×10^{-10} to 2×10^{-9} A/μm), while I_{ON} is estimated to remain largely unchanged. These results also correspond well with the TCAD-simulated pMOS having the full D_{IT} profile from Fig. 4.11(a) where I_{OFF} is increased by a factor 44× to 8.8×10^{-9} A/μm and I_{ON} is degraded only 1.6 %. The data used in this prediction were also indicated on the contour plots in Fig. 4.10(a) for reference. As a result, the proposed method can be used as a means to quickly assess the degradation in Ge MOSFET performance by an arbitrary D_{IT} profile, without the need for time consuming TCAD simulations.

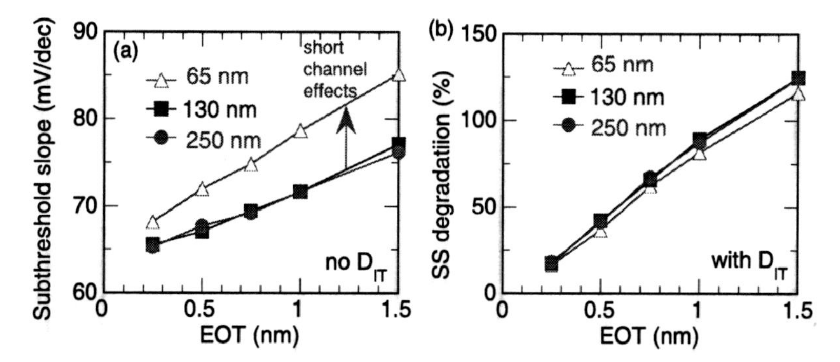

Fig. 4.12 (**a**) TCAD-simulated subthreshold slope in the absence of interface traps, as a function of equivalent oxide thickness *EOT*, for Ge pMOSFETs with different gate lengths. Note the higher *SS* for the 65 nm devices, due to short channel effects. (**b**) Relative degradation of *SS* due to the box-like D_{IT} profile (Fig. 4.9(a)) as a function of *EOT* showing that although the initial *SS* is higher for the 65 nm pFET due to short channel effects, the relative degradation is gate-length independent

4.4.3.3 *EOT*- and L_G-Dependence

So far, most of the analysis in this work was done for a gate length L_G of 250 nm and an equivalent oxide thickness (*EOT*) of 1 nm. Extending the proposed methodology to MOSFETs with different L_G (especially in the presence of short channel effects) and *EOT* (scaled MOSFETs) would increase its practical applicability. To address this problem, Ge pMOSFETs were simulated with the D_{IT} profile from Fig. 4.9(a) and compared to their D_{IT}-free counterparts. The subthreshold slope of these pFETs without any D_{IT} is given in Fig. 4.12(a), for different *EOT* values. As expected, *SS* increases for larger *EOT*-values. Also, while the 130 and 250 nm show an almost identical *EOT*–SS relationship, short channel effects in the 65 nm pFET give rise to a higher *SS*. However, the relative degradation due to interface traps for these three L_G is still the same. In other words, degradation due to D_{IT} is additive to short channel effects (i.e. the observed *SS* increase due to short channel effects does not affect the relative increase due to traps, Fig. 4.12(b)).

Consequently, the presented model can be used to study the degradation in devices of various sizes, as the relative degradation is independent of the gate length. This conclusion was already qualitatively obtained in the discussion of Fig. 4.8, where uniform D_{IT} distributions had very similar effects on long (250 nm) and short (65 nm) pMOSFETs. Note also the linear relationship between the *SS* degradation due to D_{IT} and *EOT* in Fig. 4.12(b): as a result, the methodology can be applied to devices with different *EOT* by linearly scaling the *SS* degradation, relative to the reference devices with $EOT = 1$ nm in this work. This linear dependency of the relative degradation with *EOT* indicates that transistors with smaller *EOT* are impacted less by a given concentration of interface traps. In other words: as transistor scaling continues and the *EOT* is reduced further, higher interface trap concentrations can be tolerated.

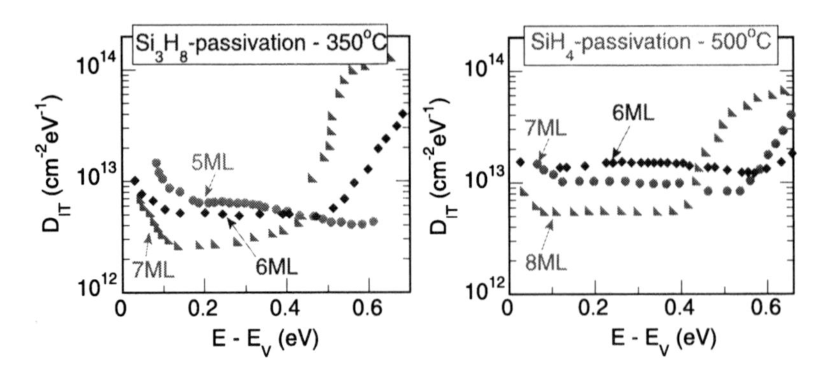

Fig. 4.13 Interface trap density profile extracted using the full conductance method for two Si–passivation techniques and different thicknesses of the deposited Si passivation layer (in monolayers—ML) [96]

4.4.4 Experimental Verification

To allow for experimental verification, Ge pMOSFETs were fabricated using different Si passivation schemes, varying the Si growth precursor (SiH_4 or Si_3H_8) and the deposited Si thickness. For each of these different passivation recipes, the interface trap spectrum was measured on a typical device using the full conductance method [96]. These profiles are given in Fig. 4.13. Using the method explained in this work, the expected subthreshold slope was calculated. These model predictions can then be compared to the measured subthreshold slope on the actual devices (Fig. 4.14). Agreement was found between the measurements and the model predictions for both passivation recipes, although the scatter on the measurements is quite large for the Si_3H_8 passivation. The observed higher *SS* in the case of the SiH_4 passivation recipe (compared to Si_3H_8) is also in line with the model and linked to higher D_{IT} values near the valence band using this precursor gas. A similar comparison for I_{OFF} is not included since this problem is equivalent to the *SS* comparison, provided I_{OFF} is dominated by subthreshold leakage. In the other case, where I_{OFF} is dominated by junction leakage, no dependency on D_{IT} is expected. The match with experimental data also proves that the main influence of interface traps on MOSFET sub-V_T behavior is through electrostatic effects.

For further experimental verification, the I_{ON} was measured on the same devices, together with *CET* (Capacitance Equivalent Thickness, from C-V measurements, [99]) and high field mobility μ on the SiH_4-passivated samples from [100]. Using the 6 ML samples as a reference, the relative changes in I_{ON}, *CET* and μ are given in Fig. 4.15. For the 7 ML-sample, I_{ON} is observed to increase (+6.3 %), resulting from a *CET*-effect (the *CET* increases by 1.5 %, giving a 1.5 % drop in C_{OX} and I_{ON}) and an apparent μ-effect (μ increases by 8.1 %). However, this observed μ-effect (as measured) can be impacted severely by interface traps when measured with the classical split-CV method: the measured inversion charge $Q_{inv\text{-}measured}$ is equal to the actual Q_{INV} plus the charge captured in fast-responding interface

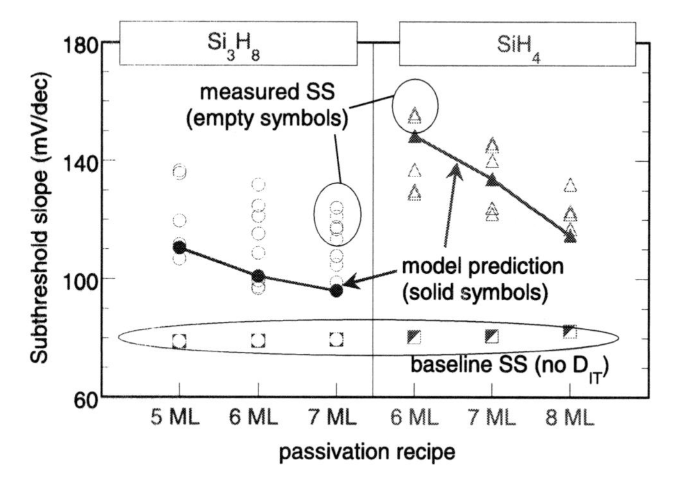

Fig. 4.14 Measured subthreshold slope for the different passivation schemes from Fig. 4.13, as compared to the model's prediction using the corresponding extracted D_{IT} profiles, on $L_G = 250$ nm Ge pMOSFETs with a measured *CET* ranging from 1.45 to 1.73 nm

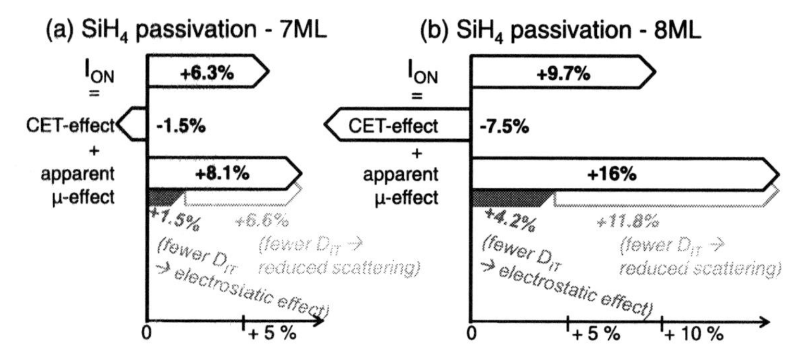

Fig. 4.15 Relative change in I_{ON} for the 7 and 8 ML SiH_4 passivated samples, comprised of a *CET*-effect and an apparent mobility effect. The mobility effect is further split in a component (1) due to electrostatic effects of interface traps and a component (2) accounting for extra scattering. Note that for both recipes, the scattering component is dominant in determining the measured mobility ($L_G = 250$ nm)

states Q_{SS}. At room temperature, this includes nearly all traps inside the Ge bandgap can be considered fast [95]:

$$Q_{inv\text{-}measured} = Q_{INV} + Q_{SS} \tag{4.6}$$

When measured $Q_{inv,measured}$ is then used to calculated channel mobility μ, this quantity is artificially lowered. Such a misinterpretation stems from the fact that the immobile charge Q_{SS} is treated in the same way as the true inversion charge Q_{INV}:

$$I_{DS} \sim Q_{inv\text{-}measured}\mu \quad \Longleftrightarrow \quad \mu \sim \frac{I_{DS}}{Q_{INV} + Q_{SS}} \tag{4.7}$$

However, the method presented can be used to split this measured mobility increase in an electrostatic component (+1.5 %: less interface traps give rise to more mobile charges, artificially increasing the measured mobility) and a true scattering component (+6.6 %). Obviously, only the latter is a true effect on mobility. In other words, while μ is observed to increase by 8.1 %, 20 % of this increase is the result of just having fewer trapped charges (smaller Q_{SS}), making the μ increase due to reduced carrier scattering a bit smaller. A similar situation is observed for the 8 ML SiH_4 passivated sample. In both cases, it must be concluded that the main influence of interface traps on the drive current I_{ON} is through scattering. The pure electrostatic effect is rather small. Nevertheless, the electrostatic effect must be filtered out, in order to get meaningful mobility numbers.

4.4.5 Conclusions

In this section, a technique was presented to study the electrostatic degradation of key Ge MOSFET performance metrics (*SS*, I_{OFF}, I_{ON}), due to an arbitrary distribution of interface traps. In this technique, an arbitrary interface trap spectrum is approximated as the sum of individual box-like trap profiles. A simple linear superposition of the contributions from each individual profile yields the total degradation. This approach was verified with TCAD simulations and experimental data by combining its prediction with measurements on Ge pMOSFETs fabricated with different passivation recipes and extracted interface trap spectra using the full conductance method. The relative *SS* degradation due to interface traps was shown to be independent of gate length, even in the presence of short channel effects in scaled devices. Additionally, a linear dependency of the relative degradation with *EOT* is observed, indicating that transistor performance metrics are impacted less by a given concentration of interface traps, as the *EOT* was reduced further. Finally, MOSFET drive current was shown to be degraded in the presence of interface traps, mainly through additional scattering in the channel, while the traps' electrostatic effect is rather small.

4.5 Summary and Conclusions

In this chapter, a TCAD device simulator was extended to allow electrical simulations of sub-100-nm Ge pMOSFETs. Firstly, parameters for mobility models in Ge were provided. A model capturing impurity scattering was calibrated based on literature data and electrical measurements on Ge pMOSFETs with L_G down to 70 nm. This allowed including mobility reduction at high lateral and high vertical electric fields. Secondly, generation-recombination in Ge was investigated on diodes, yielding a set of parameters for Shockley-Read-Hall recombination and junction leakage through trap-assisted and Band-to-Band tunneling processes.

Using this set of TCAD models, electrical simulations of Ge pMOSFETs with L_G ranging from 70 nm to 1 μm were found to be in good agreement with measured I–V curves. Specifically, typical transistor performance metrics (I_{ON}, I_{OFF}, *DIBL*, drain-to-substrate leakage, ...) on simulated pFETs were within 5–10 % of the experimental values. The electrical effect of minor changes in the processing flow (altered conditions for the halo and extension implant steps) was also accurately predicted.

The calibrated ion implant simulation tool discussed in the previous chapter, combined with the extended device simulator presented in the current one constitute a virtual processing line. Complementing the experimental work, this TCAD combination allows optimizing and predicting the performance of new, scaled germanium based devices.

Building on these TCAD capabilities, a methodology was presented allowing to study and predict the effect of interface traps in Ge technology on transistor performance. Considering an arbitrary energy distribution of interface states inside the bandgap, the resulting relative electrostatic degradation (on *SS*, I_{ON}, I_{OFF}) can be estimated using the described method. As such, time consuming electrical TCAD simulations can be avoided. The proposed approach has been verified by TCAD simulations and experimental data on different Ge pMOSFET devices with varying passivation recipes. On these devices, the relative degradation of *SS* and I_{OFF} was reproduced based solely on the interface trap spectrum, as obtained with the full conductance method. A linear dependence of the relative degradation with *EOT* was observed, indicating that transistor performance metrics *SS* and I_{OFF} are impacted less by a given concentration of interface traps if *EOT* is reduced. In other words, higher D_{IT} concentrations can be tolerated in transistors with thin *EOT* (keeping *SS*, I_{OFF} fixed).

Finally, the impact of interface traps on MOSFET drive current was investigated. In Ge MOSFETs with SiH_4 passivation, it was found that the electrostatic degradation (due to charging/decharging of traps) accounts for about 20 % of the observed change in drive current I_{ON}. Additional scattering processes (reducing carrier mobility in the channel) were found to be dominant.

Appendix

A.1 TCAD Model Parameters

This appendix contains the model parameters used in this work. Each time, both the name of physical phenomenon and the name of the parameter set is given (e.g. SRH-recombination is modeled using the *Scharfetter* dataset in Sentaurus Device) together with the default silicon values. If different values apply for electrons and holes, both values are given in this order, separated by a comma. Note that the definition of these parameters below applies solely to their specific implementation in Sentaurus Device. The literature references on which these models are based may use slightly different definitions. For this reason, the parameters below are not included in the general list of symbols of this book.

A.2 Recombination

SRH-recombination (*Scharfetter*)

Parameter	Units	Si (e, h)	Ge (e, h)
τ_{min}	s	0, 0	0, 0
τ_{max}	s	$1 \times 10^{-5}, 3 \times 10^{-6}$	$4 \times 10^{-5}, 4 \times 10^{-5}$
N_{ref}	cm^{-3}	$10^{16}, 10^{16}$	$10^{14}, 10^{14}$
γ	–	1, 1	0.85, 0.85
T_{α}	–	$-1.5, -1.5$	$-1.5, -1.5$
T_{coeff}	–	2.55, 2.55	2.55, 2.55
E_{trap}	eV	0.0, 0.0	0.0, 0.0

Note (1)—E_{trap} refers to the SRH reference trap energy w.r.t mid-bandgap (e.g. $E_{trap} = 0.0$ corresponds to $E_V + 0.33$ eV in Germanium)
Note (2)—τ_{max} was taken from [47] and then decreased slightly to correspond with the leakage measurements on our Ge p+/n diodes
Note (3)—The temperature dependence for Ge was not investigated. Instead, Si defaults are still used. Further research is required for this dependency

TAT (*HurckxTrapAssistedTunneling*)

Parameter	Units	Si (e, h)	Ge (e, h)
m_t	–	0.5, 0.5	0.12, 0.34

BTBT (*Band2BandTunneling*)

Parameter	Units	Si	Ge
A	$cm\,s^{-1}\,V^{-2}$	8.977×10^{20}	8.977×10^{20}
B	$eV^{-3/2}\,V\,cm^{-1}$	2.147×10^{7}	1.6×10^{7}

A.3 Mobility

Phonon Scattering (*ConstantMobility*)

Parameter	Units	Si (e, h)	Ge (e, h)
μ_{max}	$cm^2\,V^{-1}\,s^{-1}$	1417, 470.5	3900, 1900
exponent	–	2.5, 2.2	2.5, 2.2

Impurity Scattering (*DopingDependence*)

Parameter	Units	Si (e, h)	Ge (e, h)
μ_{min1}	cm^2/Vs	52.2, 44.9	60, 60
μ_{min2}	cm^2/Vs	52.2, 0.0	0, 0
μ_1	cm^2/Vs	43.4, 29	20, 40
P_c	cm^{-3}	$0, 9.23 \times 10^{16}$	$10^{17}, 9.23 \times 10^{16}$
C_r	cm^{-3}	$9.68 \times 10^{16}, 2.23 \times 10^{17}$	$8 \times 10^{16}, 2 \times 10^{17}$
C_s	cm^{-3}	$3.34 \times 10^{20}, 6.10 \times 10^{20}$	$3.43 \times 10^{20}, 10^{20}$
α	–	0.68, 0.719	0.55, 0.55
β	–	2.0, 2.0	2.0, 2.0

High Lateral Field Mobility (*HighFieldDependence*)

Parameter	Units	Si (e, h)	Ge (e, h)
v_{sat0}	cm/s	$1.07 \times 10^7, 8.37 \times 10^6$	$8 \times 10^6, 6 \times 10^6$

High Transversal Field Mobility (*EnormalDependence (holes only)*)

Parameter	Units	Si	Ge
B	cm/s	9.925×10^6	1.993×10^5
C	$cm^{5/3}\, V^{-2/3}\, s^{-1}$	2.947×10^3	4.875×10^3
N_0	cm^{-3}	1	1
λ	–	0.0317	0.0317
k	–	1	1
δ	cm^2/Vs	2.0546×10^{14}	1.705×10^{11}
A	–	2	1.5
$\alpha_\perp$	cm^3	0	0
N_1	cm^{-3}	1	1
ν	–	1	1
η	$V^2\, cm^{-1}\, s^{-1}$	2.0546×10^{30}	2.0546×10^{30}
l_{crit}	cm	10^{-6}	10^{-6}

Chapter 5
Investigation of Quantum Well Transistors for Scaled Technologies

In this chapter, after discussing scaling issues in bulk silicon and bulk germanium MOSFET technology, heterostructure confinement is investigated as a means to enhance MOSFET scalability. The Implant-Free Quantum Well FET is introduced and its performance analyzed using TCAD simulations.

5.1 Introduction

Germanium has emerged as an exciting alternative material for scaled logic applications. The most obvious strength of this material is undoubtedly its higher bulk carrier mobility, compared to that of silicon [18], since Ge offers the highest bulk hole mobility of all known semiconductors. Although the reduction of L_G makes the transport mechanism change from diffusive [94] to ballistic transport [89, 108], the drive current I_{ON} has been claimed to depend on the bulk carrier mobility. On the other hand, the static power dissipation of integrated circuits (leakage) is becoming an increasingly important issue [144]. Germanium's smaller band gap, as shown in the previous chapters, will lead to more junction leakage through Trap-Assisted Tunneling and Band-to-Band tunneling mechanisms. As such, a scaled VLSI technology integrating Ge-based transistors will have to provide a boost in active performance, keeping leakage issues under control.

At the same time, planar bulk silicon technology, which has been the workhorse of the IC-industry for the past decades is also running out of steam and has required many material innovations in addition to the classical area scaling to provide the required performance [20].

Considering this, the first goal of this chapter is to investigate some of the scaling issues in bulk MOFSET technology. Considering bulk Ge pFET technology, drain-to-bulk junction leakage will be studied, focussing on the conditions that may allow a bulk Ge pFET technology to meet the I_{OFF} specifications. Considering bulk Si pFET technology, the electrical effects of imposing fixed source/drain junctions and a maximum halo implant dose are assessed. Especially the short channel control and scalability are studied.

G. Hellings, K. De Meyer, *High Mobility and Quantum Well Transistors*, Springer Series in Advanced Microelectronics 42, DOI 10.1007/978-94-007-6340-1_5,

The second section of this chapter will study heterostructure confinement as a means to obtain a transistor structure with enhanced scalability, compared to the bulk technologies discussed in the first section. The Implant-Free Quantum Well FET is introduced and its performance analyzed for gate lengths down to 16 nm. Using the Si/SiGe material system as an example-case, the link between heterostructure confinement and (the reduction of) short channel effects is explored. To this end, the TCAD simulation models developed in the previous chapters are used.

Finally, the Implant-Free Quantum Well nFET will be introduced starting from the High Electron Mobility Transistor (HEMT) on which its original design is based [52, 54].

5.2 Motivation—Scalability Issues in Bulk MOSFET Technologies

This section will investigate two scaling issues in bulk MOSFET technologies. One issue for Ge technologies is the high drain-to-bulk junction leakage, due to Ge's small bandgap E_G of 0.66 eV. While in Si technologies junction leakage is typically only a minor concern, it may become the dominant source of power consumption in a Ge technology. This concern has led to several studies [38, 39, 56] on junction leakage mechanisms in Ge, notably Trap Assisted Tunneling (TAT) and Band-to-Band Tunneling (BTBT), and their effect on transistor characteristics [79–81, 93]. Experimentally, drain-to-bulk leakage currents in excess of 1 μA/μm have been consistently observed in the 70 nm Ge pMOSFET technology, discussed earlier in Chap. 4. This value is clearly too high considering that total OFF-state leakage current densities below 100 nA/μm are typically required for high performance logic applications [75].

Secondly, the effect of imposing fixed source/drain junctions and a maximum halo implant dose on scaling a Si pFET is investigated. The goal of this exercise is to show how exactly L_G scaling would affect transistor characteristics in the hypothetical scenario where source/drain junctions and halo doping cannot be changed anymore without undesired side effects. Obviously, the focus is on qualitative conclusions regarding the nature of the short channel effects and their effect on transistor I_D–V_G characteristics.

5.2.1 *Drain Extension Leakage in a 65 nm Bulk Germanium pMOS Technology*

The total drain-to-bulk junction leakage can be decoupled in three components (Fig. 5.1, [39]). The area leakage is generated under the HDD implants and scales directly with the area of the drain region. The isolation leakage is generated at the

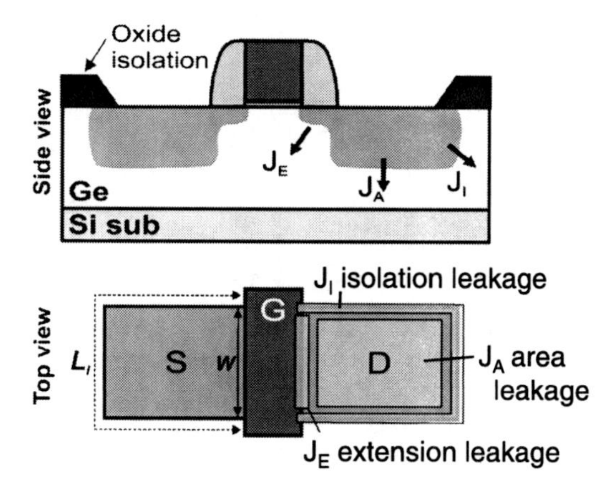

Fig. 5.1 Schematic cross section and (*bottom*) top view of a bulk germanium transistor, indicating three contributions to the junction leakage: area leakage, isolation leakage and extension leakage, [39]

interface between the drain region and the transistor's isolation. Finally, the extension leakage is generated under the transistor's spacer regions and scales with the transistor width W.

The first two components have been intensively studied on diodes in [38] and were found to remain well below the power-density specifications for high-performance applications. The last component—extension leakage—was quantified experimentally in [39] and found to be the dominant junction leakage component in short-channel Ge pFETs: the halo implants increase the electric field at the drain side, leading to enhanced TAT and BTBT leakage. Furthermore, a trade-off was found between short channel control on the one hand and drain leakage on the other: changing the halo or extension implant conditions will provide better short channel control only at the expense of increased leakage. An example of this trade-off is shown in Fig. 5.2, where the extension leakage is shown as a function of supply voltage V_{DD} and implanted halo dose. Lowering the 80 keV As halo implant dose from 5×10^{13} cm^{-2} to 3.5×10^{13} cm^{-2} succeeds in reducing the extension leakage, but only at the expense of worsening short channel behavior (DIBL increases from 185 mV/V to 280 mV/V). Increasing the halo implant dose to 6.5×10^{13} cm^{-2} only mildly improves short channel control (DIBL is reduced to 150 mV/V from 185 mV/V for $L_G = 70$ nm), while the sharper junction (larger electric field) yields increased extension leakage.

For the 65 nm germanium pFET technology (As halo, 5×10^{13} cm^{-2}) with a DIBL value of 185 mV/V, extension leakage current densities below 100 nA/µm could only be obtained for a supply voltage equal to or lower than 0.7 V (at room temperature). Clearly smaller gate lengths would require heavier halo doses to keep short-channel effects under control, requiring an even smaller supply voltage to keep the extension leakage under control. With these limitations in mind, it seems unlikely that a bulk Ge MOSFET technology would be able to deliver the promised performance in future technology nodes ($L_G < 20$ nm). As such, a different transistor structure will be required, designed to avoid excessive junction leakage.

In bulk silicon MOSFET technology, the junction leakage is less of a problem due to the larger band gap of this material ($E_G = 1.12$ eV), comparing transistors

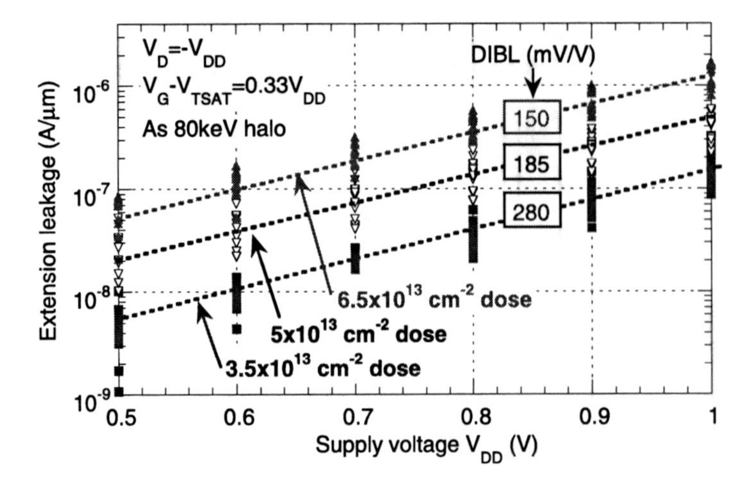

Fig. 5.2 Extension leakage as a function of supply voltage V_{DD} for the 65 nm Ge pFET technology. The implanted halo dose was varied starting from the reference value of 5×10^{13} cm^{-2}. Notice that the maximum allowed supply voltage (keeping the extension leakage below 100 nA/μm) is lower for higher halo doses, [39]

with the same gate length. However, silicon is not at all immune to the trade-off discussed above. The larger band gap will merely postpone this issue to a smaller gate length. As a result, bulk silicon technology is believed to be confronted with similar tradeoffs at more advanced technology nodes.

5.2.2 *Short Channel Effects in Bulk Si Technologies*

As transistor gate length scaling continues, control over short channel effects becomes increasingly difficult in planar bulk CMOS technologies. The classical recipe to mitigate this effect has been to increase channel (or halo) doping. However, this gives rise to undesired effects such as dopant induced fluctuations in threshold voltage and degradation in channel mobility through additional impurity scattering. Considering also the junction leakage issues discussed for Ge in the previous paragraphs, increased halo doping level will—at one point in time—not provide improved transistor characteristics.

In order to illustrate that situation, the following paragraphs will discuss the scalability of a bulk Si pMOS technology keeping the S/D junctions and halo-implant condition constant. The goal of this exercise is to show how exactly L_G scaling would affect transistor characteristics in the (hypothetical) scenario where source/drain junctions and halo doping cannot be changed anymore without undesired side effects. Note that our specific junction/halo conditions are not meant to represent fully optimized ultimate junction conditions for bulk silicon technology. Needless to say, quantitative conclusions of this study will depend heavily on those conditions. The mechanism behind short channel effects has already been discussed extensively in literature. However, specific aspects will be explored in detail, partly serving as an introduction to the next section.

5.2.2.1 Bulk MOSFET Structure and TCAD Models

The bulk Si MOSFET structure used in this section is represented schematically in Fig. 5.6(a) and can be considered a classical short-channel MOSFET. In the next paragraphs, changes on the L_G parameter will be investigated, other parameters will remain unchanged. Note the following important points:

- Source and drain extension active doping level and junction depth are fixed for all L_G at 1×10^{20} cm^{-3} and 20 nm respectively. Both extend 3 nm under the gate (underlap), creating a low-resistive contact between the channel and the S/D junction.
- Halo implant energy and dose are also kept constant as a function of L_G, using a phosphorus implant at an energy of 10 keV 3×10^{13} cm^{-2}. Using these conditions, on blanket wafers, a peak halo dose of 1×10^{19} cm^{-3} would be reached at a depth of 17 nm (the straggle for the as-implanted profile is about 25 nm).
- All other transistor parameters are kept constant as a function of gate length scaling ($EOT = 1$ nm, $N_{WELL} = 10^{17}$ cm^{-3}).

The effect of gate length scaling on these bulk Si MOSFETs is investigated using TCAD simulations (Sentaurus Device, [127]). The mobility models for Si were mentioned already in the previous chapter and include bulk mobility, impurity scattering, velocity saturation at high lateral fields and acoustic phonon scattering/surface roughness at high vertical fields. Generation-recombination models were not included for clarity, since the goal of this exercise is to investigate transistor scalability for short gate lengths. The MOSFETs are characterized at a supply voltage of $V_{DD} = 1$ V. I_{ON} and I_{OFF} are extracted at a fixed offset voltage from the saturation threshold voltage $V_{T,sat}$ ($-0.7 \times V_{DD}$ and $0.3 \times V_{DD}$ respectively).

5.2.2.2 Short Channel Control

In Fig. 5.3(a), the simulated MOSFET drain current I_D is plotted as a function of V_G, for L_G ranging from 90 nm, down to 16 nm. For $L_G \geq 45$ nm, a steep subthreshold regime can be observed i.e. reducing the gate length has a positive effect on I_{ON}, with little or no impact on I_{OFF}. A rather steep subthreshold slope of 90 mV/V and DIBL values close to or below 100 mV/V (Fig. 5.3(b)), illustrate the good gate control. However, scaling L_G further to 32 and 22 nm, has a pronounced impact on the transistor characteristics. Notably, *DIBL* quickly increases to 186 ($L_G = 32$ nm) and 631 mV/V ($L_G = 22$ nm). Ultimately, noteworthy transistor action almost completely vanishes in the $L_G \approx 16$ nm device, with an I_{ON}/I_{OFF} ratio close to 10.

In order to explain this behavior, the simulated OFF-state current density is plotted in Fig. 5.3(c), (d) for the $L_G = 22$ nm and the $L_G = 65$ nm pFET. The latter

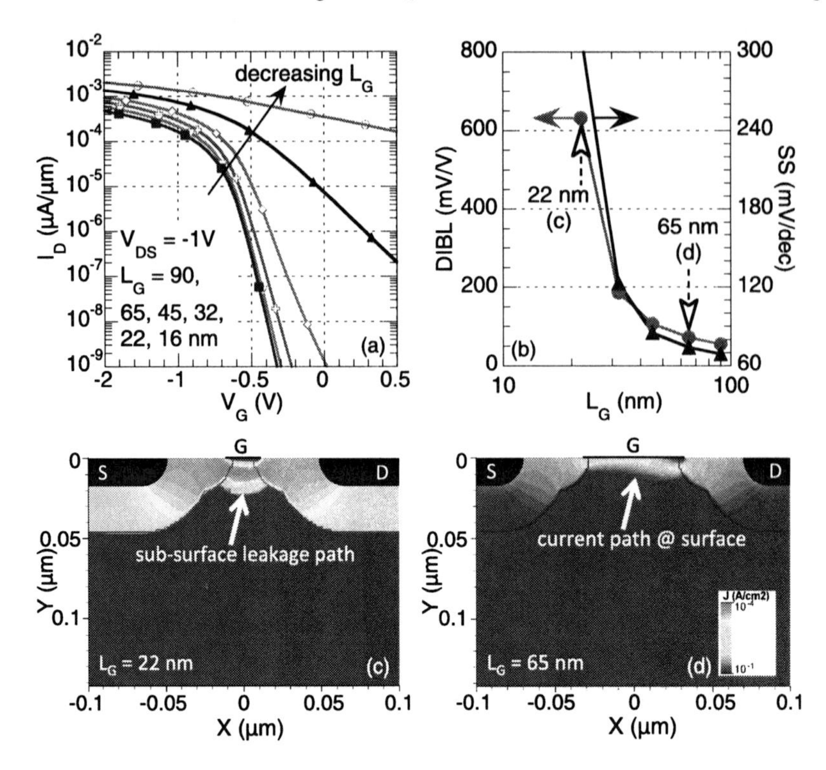

Fig. 5.3 (**a**) Simulated I_D–V_G curves for a bulk Si pFET, fixing source/drain structure and halo. Notice the pronounced short channel effects for L_G below 45 nm. (**b**) *DIBL* and *SS* as a function of gate length for these pFETs. TCAD-simulated OFF-state current density for the $L_G = 22$ (**c**) and $L_G = 65$ nm (**d**) pFET, showing the transition from a surface-dominated current path in long channel FETs towards the buried, sub-surface path in the smaller gate lengths

device has a small off-state leakage current, flowing at the Si/SiO_2 interface. In the $L_G = 22$ nm pFET, a sub-surface OFF-state leakage path can be observed connecting source and drain about 10 nm below the Si/SiO_2 interface. Because this buried current path is relatively far from the gate, it cannot easily be modulated by the gate voltage, resulting in a strongly degraded subthreshold slope.

Examining this leakage mechanism in detail, 3D-plots of the above transistors (in OFF-state) were made with L_G ranging from 90 nm to 22 nm in Fig. 5.4. The z-axis (pointing downwards) shows the valence band energy E_V as a function of the x and y coordinates, relative to the Fermi energy level E_F at the source. In this manner, a 3-dimensional surface of the energy barrier charge carriers (i.e. holes) have to overcome when flowing from source to drain is generated. For long channels, source and drain are separated by an energy barrier of 300 to 400 meV (green on the z-axis color scale). Only few holes will be able to overcome this energy barrier and consequently only a very small current will flow from source to drain (i.e. I_{OFF} will be low). Secondly, for long channels, the lowest-energy path from source to drain is located at the semiconductor/dielectric interface. As a result, most of the I_{OFF}

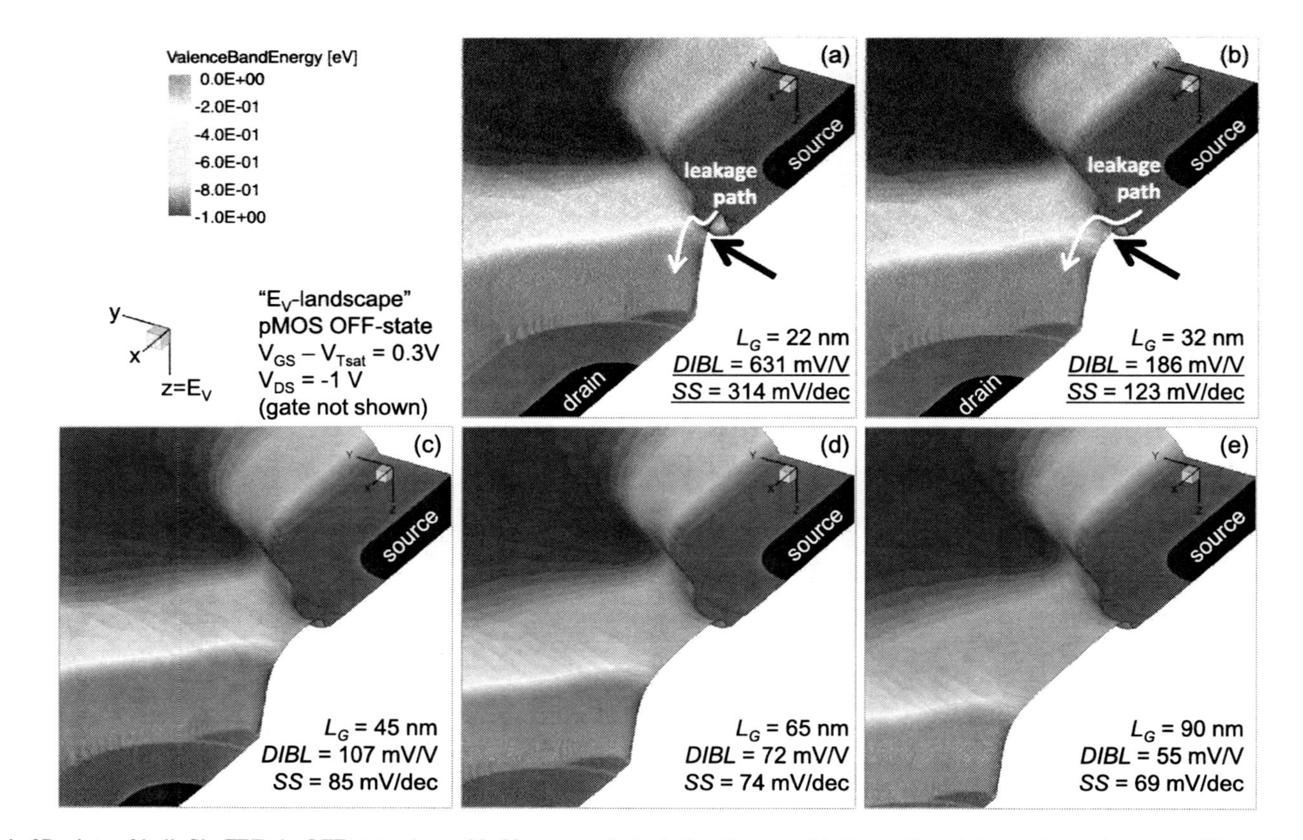

Fig. 5.4 3D-plots of bulk Si pFETs in OFF-state, $L_G = 90$–22 nm. z-axis (pointing downwards) shows E_V relative to E_F at the source. Notice the reduced energy barrier (source to drain) for short L_G (Color figure online)

current will flow at this interface where it can be efficiently modulated by the gate electrode, giving rise to a good subthreshold slope *SS* and low *DIBL* values. The OFF-state current flow at the surface was shown previously in Fig. 5.3(d) for the $L_G = 65$ nm pFET.

For the smaller pFETs (e.g. the $L_G = 22$ nm device, Fig. 5.4(a)), the situation is different. Source and drain are no longer separated by a large energy barrier. Instead a saddle point can be observed in the 3D energy surface, several nm below the semiconductor/dielectric interface. As a result, more holes will be able to overcome this smaller energy barrier, resulting in a high current flow from source to drain (i.e. I_{OFF} will be rather large). Secondly, as the lowest-energy path from source to drain is located below the semiconductor/dielectric interface, most of the current will follow this sub-surface leakage path, where it cannot be efficiently modulated by the (more distant) gate electrode. Consequently, *SS* and *DIBL* values will be rather high. Note that the OFF-state current density for the $L_G = 22$ nm pFET was also shown previously in Fig. 5.3(c): a clear sub-surface current path is visible, 8 nm below the gate dielectric.

In conclusion, keeping the S/D junctions and halo implant condition constant while scaling L_G has a detrimental effect on device characteristics. Notably, sub-surface leakage path ultimately causes a sharp increase in *DIBL* and *SS*. Consequently, when s/d junctions and halo implant conditions cannot be scaled further, a non-bulk architecture will be required, designed to avoid this type of leakage. Multi-gate FETs and Silicon-On-Insulator (SOI) FETs have been developed to provide better scalability by eliminating such sub-surface current paths.

5.2.3 Conclusions

In this section, two scalability studies were performed. In the first one, focusing on bulk germanium MOSFET technology, drain-to-bulk junction leakage through SRH, TAT and BTBT mechanisms was investigated. A trade off between good short channel control and low junction leakage was found for a 65 nm germanium pFET technology. Extension leakage current densities below the ITRS spec of 100 nA/µm could only be obtained for supply voltages of 0.7 V or lower (keeping the reference halo implant). Because smaller gate lengths would require heavier halo implants, it seems unlikely that a bulk Ge MOSFET technology would be well suited for future technology nodes ($L_G < 20$ nm). Rather, different transistor architectures, designed to avoid excessive junction leakage, would be beneficial.

The second study investigated the effect of imposing fixed source/drain junctions and a maximum halo implant dose on a scaling silicon pFET. As expected, this resulted in severely degraded transistor I_D–V_G characteristics for smaller gate length devices. More precisely, a sub-surface leakage path was shown to appear using TCAD simulations, causing high *DIBL* and *SS* values. Consequently, when increasing the halo dose in bulk Si technology is no longer an option, alternative transistor architectures are required, different from the bulk Si or bulk Ge structures investigated so far. A key feature of such an architecture should be to inhibit leakage current mechanisms discussed in this section.

5.3 Towards a Scalable Transistor Architecture

In the previous section, it became clear that scalable transistor architectures are required, avoiding the various leakage mechanisms encountered in standard bulk Si FETs. Alternative and more scalable transistor architectures are being developed, such as multi-gate [28] and Silicon-On-Insulator FETs (SOI, [27]). The common feature of these structures is superior electrostatic gate control, which is achieved by confining (restricting) the charge carriers to a well-defined volume (i.e. the fin in multi-gate FETs or the thin Si layer in SOI-based FETs). While successful in terms of scalability, these alternatives both come with an increased wafer cost (SOI-based FETs) or process complexity (vertical dimension in multi-gate FETs).

In this section, a third group of devices will be introduced, which confines the charge carriers to a Quantum Well (QW), formed by a heterostructure. This concept has been experimentally demonstrated in a SiGe Implant-Free Quantum Well pFET, showing excellent short channel control down to $L_G = 30$ nm. While experimental results will be covered in more detail in the next chapter, this section will focus on the concept of *heterostructure confinement*. This concept, and its implementation into the SiGe Implant-Free Quantum Well FET will be explained using TCAD simulations.

5.3.1 Heterostructures: Fermi Level Continuity

One feature of heterostructures will prove particularly useful in this section, namely that (in equilibrium condition) the Fermi energy level E_F is continuous at the interface between two materials. This is shown schematically in Fig. 5.5: two materials (A and B) are brought into contact. Material A has a smaller band gap E_G than material B, resulting in band offsets ΔE_V and ΔE_C for the valence and conduction bands respectively. The Fermi energy level continuity across the hetero-interface causes the electron concentrations in both materials to be different. Indeed, considering that the electron concentration in the conduction band E_C is given by [138]:

$$n = N_C e^{\frac{E_F - E_C}{k_b T}},$$

the electron concentration in both materials is different by a factor F

$$F = e^{\frac{\Delta E_C}{k_b T}}, \tag{5.1}$$

assuming N_C for both materials is identical. This factor F quickly increases, yielding a few orders of magnitude difference in electron concentrations in the adjacent materials.

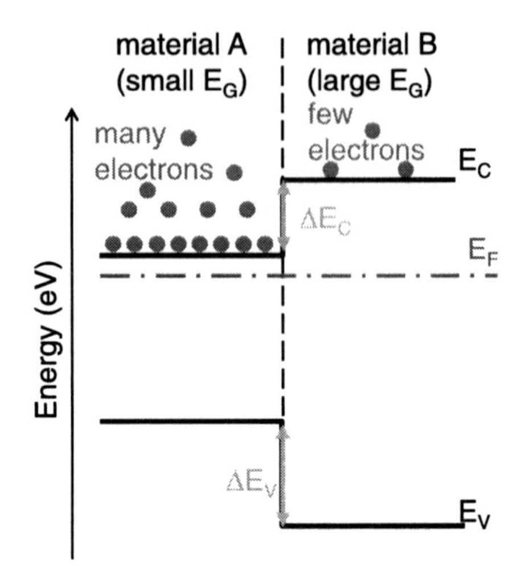

Fig. 5.5 Schematic representation of a heterostructure interface in equilibrium. The continuity of the Fermi energy E_F can cause a difference in conduction band electron concentrations for both materials

5.3.2 A Case Study: SiGe FETs

The effects of integrating heterostructure confinement into transistor architectures will be discussed. Using the Si/SiGe material system as an example-case, the link between heterostructure confinement and (the reduction of) short channel effects is explored. The Implant-Free Quantum Well FET is introduced and its performance analyzed for gate lengths down to 16 nm.

5.3.2.1 Transistor Structures and Modeling

This comparative study is centered around two transistor structures: the bulk Si pFET and the SiGe Implant-Free Quantum Well (IFQW) pFET. The latter is fundamentally different from the bulk FET in two key aspects: (1) no halo and junction implants are performed. Instead, an epitaxial growth process is used to fabricate source/drain areas (*Implant-Free*) and (2) a Si/SiGe/Si quantum well is used to confine the charge carriers (i.e. holes) to a thin, well-defined SiGe channel (*Quantum Well*). In order to provide more understanding, two additional FET structures were added to this study. These in-between cases will allow to separate the effect of each of the above-mentioned aspects. Figure 5.6 contains a schematic drawing of the four transistor structures considered. Each of these was simulated with Sentaurus Process [128] and will now be discussed in detail. Note that the described simulated process flows may differ in some aspects from a real fabrication process. Some aspects have indeed been simplified for the sake of obtaining a more straight-forward comparison (e.g. well doping implant and anneal are not simulated, instead a uniform n-type active doping concentration is included in the starting substrate).

The *bulk Si pMOSFET* (Fig. 5.6(a)) is identical to the one described in Sect. 5.2.2.1. Starting from a bulk Si substrate with an initial n-type well doping

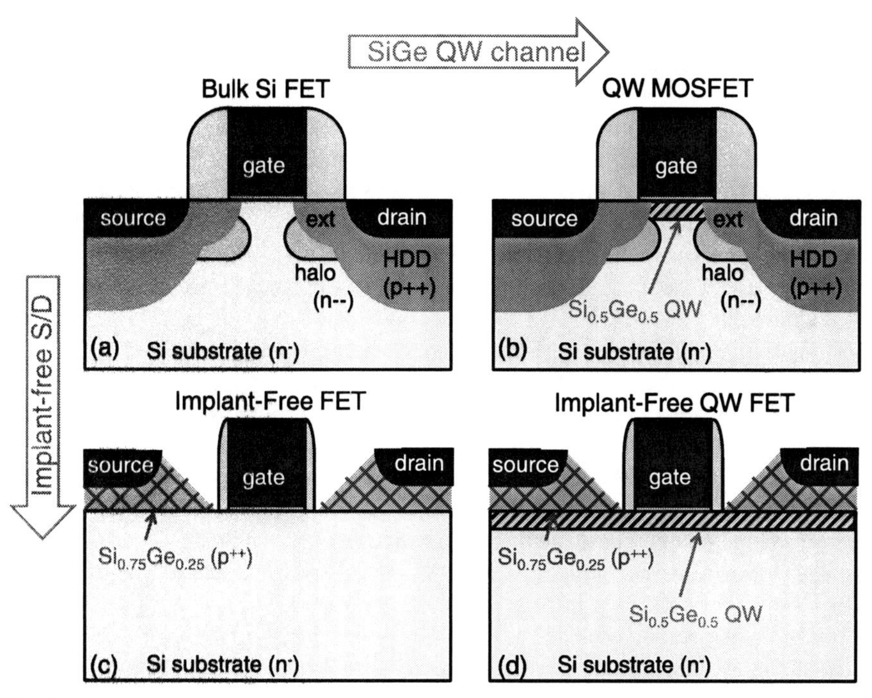

Fig. 5.6 Schematic drawing of the four transistor architectures considered in this study, centered around the bulk Si pFET (**a**) and the SiGe Implant-Free Quantum Well (IFQW) pFET, featuring a Quantum Well channel in combination with a raised source/drain morphology (**d**). The Quantum Well FET and Implant-Free FET (resp. (**b**) and (**c**)) can be considered as waypoints between (**a**) and (**d**), each of them including only one change with respect to the Si pFET

concentration of 10^{17} cm^{-3}, a gate is formed ($EOT = 1$ nm), followed by the halo implant (Phosphorus, 10 keV 3×10^{13} cm^{-2}). The source and drain extension active Boron doping level is 1×10^{20} cm^{-3} (depth = 20 nm) and extends 3 nm under the gate. Standard spacers are deposited against the gate sidewall (55 nm wide). The HDD regions have a slightly higher doping level of 2×10^{20} cm^{-3} and extend 45 nm into the Si. Finally, a silicide contact is formed, extending 17 nm into the silicon.

Fabrication of the *SiGe Implant-Free Quantum Well pFET* (Fig. 5.6(d)) also starts from a bulk Si substrate with an initial n-type well doping concentration of 10^{17} cm^{-3}. Then, 6 nm of silicon is etched out after which the Quantum Well is created by growing the following layers consecutively: 2 nm of undoped silicon, 3 nm of undoped $Si_{0.50}Ge_{0.50}$ and 1 nm of undoped Si. Because the top layer of the deposited stack is silicon, a standard gate can be formed ($EOT = 1$ nm). Immediately after gate formation, a thin offset spacer (4 nm) is deposited against the gate sidewalls. Then, raised, $Si_{0.75}Ge_{0.25}$, B-doped (1×10^{20} cm^{-3}) source and drain regions are grown epitaxially. These are then also contacted with a silicide. Note that this flow does not rely on ion implant steps to form the source and drain junctions or halo regions. Finally, $Si_{0.75}Ge_{0.25}$ is used as material for the raised source/drain regions. This allows to strain the channel, an effect which will be discussed in detail in the next chapter. In this chapter, for the sake of simplicity, strain effects are ignored.

The first in-between transistor structure is the *SiGe Quantum Well pFET* and is depicted in Fig. 5.6(b). It is identical to the bulk Si pFET, except for the inclusion of a Quantum Well channel. As such, the processing starts from a bulk Si substrate with an initial n-type well doping concentration of 10^{17} cm^{-3}. Then, 6 nm of silicon is etched out after which the Quantum Well is created by growing the following layers consecutively: 2 nm of undoped silicon, 3 nm of undoped $Si_{0.50}Ge_{0.50}$ and 1 nm of undoped Si. The remainder of the process is identical to the bulk Si pFET flow (doping profiles (junctions) are identical to those of the bulk Si pMOSFET).

The second in-between transistor structure is the *SiGe Implant-Free pFET* (Fig. 5.6(c)). Its processing flow is identical to that of the bulk Si pFET up to the gate formation. After that, it follows the implant-free processing scheme also found in the IFQW pFET. A thin offset spacer (4 nm) is deposited against the gate sidewalls. Then, raised, $Si_{0.75}Ge_{0.25}$, B-doped (1×10^{20} cm^{-3}) source and drain regions are grown epitaxially. These are then also contacted with a silicide.

These four transistor structures will be studied using Sentaurus Device [127]. The mobility models for Si include bulk mobility, impurity scattering, velocity saturation at high lateral fields and acoustic phonon scattering/surface roughness at high vertical fields. Generation-recombination models were not included to allow a clear investigation of the scalability for these transistor structures. The same models were also implemented for SiGe, using interpolated, mole-fraction dependent model parameters based on the available model set for silicon and germanium. The FETs are characterized at a supply voltage of $V_{DD} = 1$ V. I_{ON} and I_{OFF} are extracted at a fixed offset voltage from the saturation threshold voltage $V_{T,sat}$ ($0.7 \times V_{DD}$ and $0.3 \times V_{DD}$ respectively).

5.3.2.2 Electrical TCAD Simulations

In order to investigate the transistor behavior for short gate lengths, *DIBL* and *SS* were extracted as a function of L_G for all transistor architectures (Fig. 5.7). As shown before, both parameters are quickly increasing for L_G smaller than 45 nm for the bulk Si pFET. Both in-between cases (SiGe Quantum Well pFET and the Implant-Free pFET) perform somewhat better (e.g. *SS* for the $L_G = 32$ nm devices are close to or below 100 mV/dec, as compared to 123 mV/dec for the bulk Si pFET. The Implant-Free Quantum Well pFET however, exhibits superior short channel control with a *DIBL* and *SS* of 143 mV/V and 88 mV/dec respectively. For longer L_G *DIBL* and *SS* converge towards the same value for all considered transistor architectures.

Besides short channel control, the drive current I_{ON} is of course also an important parameter. Including I_{ON} in the comparison was done in the I_{ON}–I_{OFF} plot in Fig. 5.8. It can be seen that the I_{ON}–I_{OFF} data points for long channels are very close to each other ($I_{OFF} \approx 10^{-10}$ A/µm and $I_{ON} = 300$–400 µA/µm),

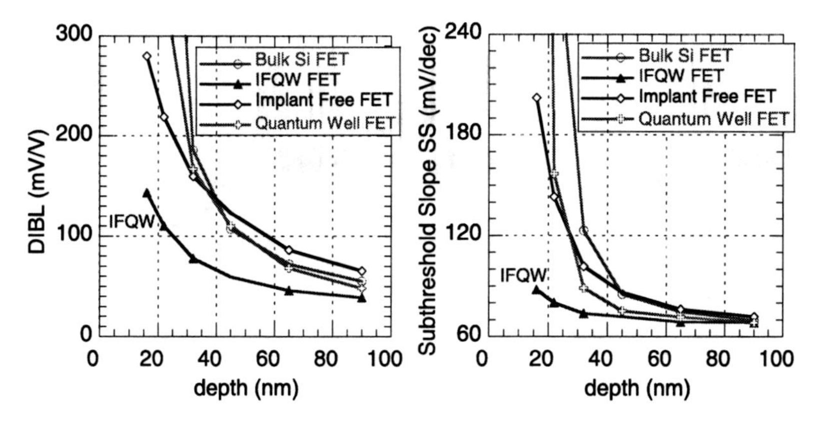

Fig. 5.7 *DIBL* and *SS* as a function of gate length for the different transistor architectures. Notice the enhanced short channel control in the Implant-Free Quantum Well pFET

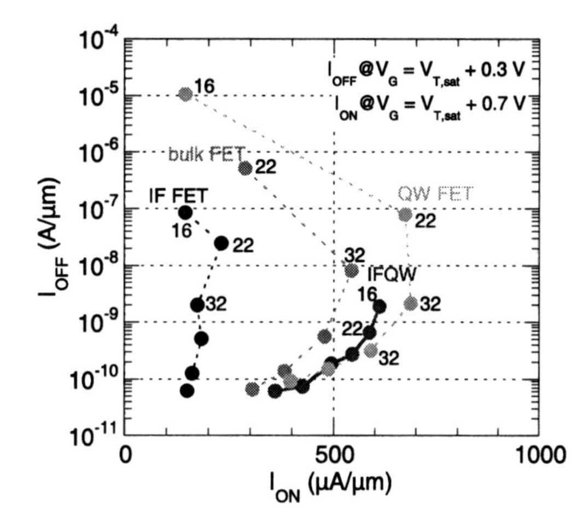

Fig. 5.8 I_{ON}–I_{OFF} plot for the different transistor architectures. Notice the "snap-back" towards *the upper left corner* in the presence of short channel-effects (e.g. $L_G = 22$ nm bulk Si pFET), occurring later for the QW-based devices, as compared to the Si pFET (L_G next to the datapoints in nm)

except for the Implant-Free FET, where I_{ON} is impacted by a rather large external resistance R_{EXT}. The situation is quite different for short channel devices however. Taking the $L_G = 22$ nm device as a reference, the effect of a more scalable transistor design becomes obvious: I_{ON} collapses for the bulk Si pFET and the Quantum Well pFET. Especially the I_{ON} for a given target I_{OFF} of 100 nA/µm, reveals the positive effect of a more scalable transistor structure on the drive current: in order to keep a fixed I_{OFF}, the V_T has to be increased, leaving a small available overdrive $(V_{G,ON} - V_T)$.

5.3.2.3 Heterostructure Confinement—Short Channel Control

In order to investigate the cause of the improved short channel control, the OFF-state current density was plotted for the transistor structures with $L_G = 22$ nm in

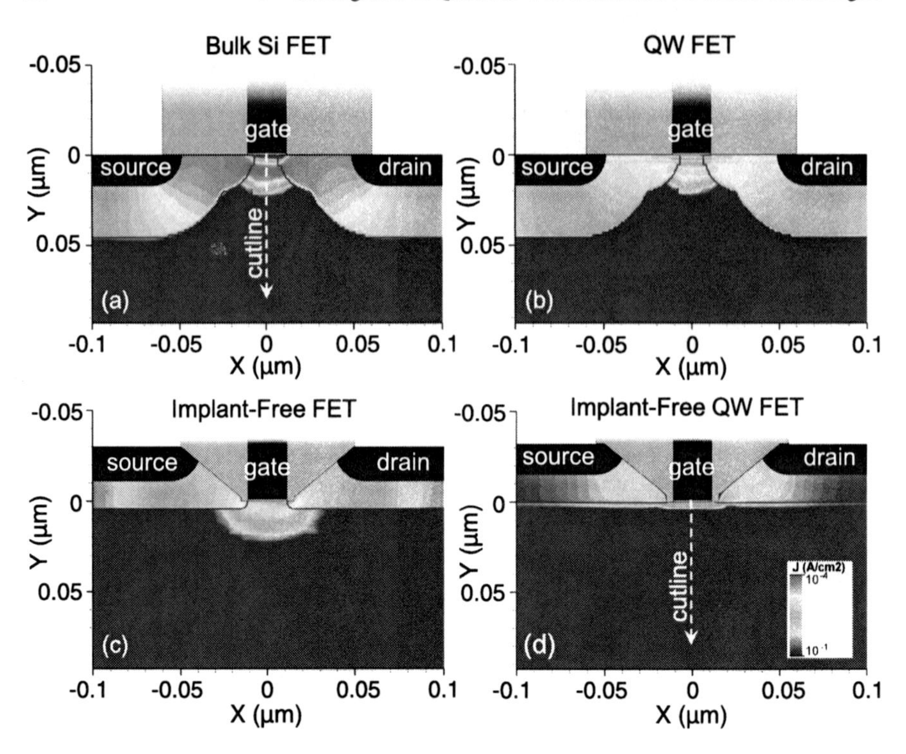

Fig. 5.9 Simulated OFF-state current density for the four transistor architectures at $L_G = 22$ nm. Notice the reduction of the sub-surface leakage path in the QW FET and the IF FET and its disappearance in the IFQW pFET, due to the valence band offset ΔE_V

Fig. 5.9. As discussed, a clear sub-surface leakage path is visible, 8 nm below the gate dielectric, in the bulk Si pFET (Fig. 5.9(a)). This is clearly the cause of the high I_{OFF} and degraded *DIBL*, *SS* observed in this device.

Observing the OFF-state current density for the Implant-Free pFET in Fig. 5.9(c), this sub-surface leakage path is clearly less pronounced, although still present. This gives an explanation for the observed difference in I_{OFF} between the IF pFET and the bulk Si pFET. The raised, epitaxially grown source/drain areas succeed in improving the short channel control, essentially by moving source and drain further apart.

In the SiGe QW pFET (Fig. 5.9(b)), the OFF-state current is also reduced, as compared to the bulk Si pFET. Here, two parallel leakage paths can be observed. At about 8 nm below the gate oxide, the same sub-surface leakage path can still be observed. Additionally, a second leakage path is observed in the SiGe Quantum Well itself. As such, introducing a QW channel into the classical junction-based MOSFET structure succeeds in reducing the leakage in the underlying Si substrate. However, the proximity of the S/D extensions still causes S/D leakage inside the QW channel. More in-depth analysis of this architecture revealed a high sensitivity to the exact gate length, source/drain morphology and halo implants. A small change

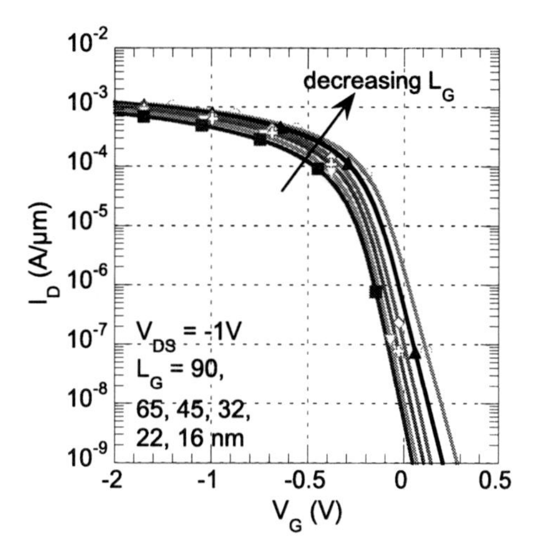

Fig. 5.10 Simulated I_D–V_G curves for the SiGe IFQW pFET. Notice the less pronounced short channel effects as compared to the Si pFET (Fig. 5.3(a)), and the rather steep SS down to $L_G = 16$ nm

in a critical parameter can already eliminate most of the sub-surface leakage; the two-dimensional proximity effect of S/D did cause the leakage in the QW itself to remain. These effects are not discussed here, although this should be kept in mind when evaluating this device experimentally.

Finally, the Implant-Free Quantum Well pFET (Fig. 5.9(d)), combines the raised source/drain structure with the QW channel. Here, the sub-surface leakage path is absent and QW channel leakage is minimal, yielding a low I_{OFF}. The enhanced short channel control also becomes clear when observing the I_D–V_G characteristics for the IFQW pFET with L_G ranging from 16 to 90 nm (Fig. 5.10): a steep SS region can be observed for all devices, without evidence of the punch-through effect observed in the bulk Si pFET (Fig. 5.3(a)).

Two additional sets of figures go into more detail on the concept of heterostructure confinement, which is responsible for providing the improved scalability. In the first set (Fig. 5.11), the OFF-state valence band energy and hole density are plotted along the vertical cutline in the middle of the channel, as indicated in Fig. 5.9. For longer gate lengths, the E_V has a maximum at the Si/dielectric interface, leading to a diminishing hole density as a function of depth (Fig. 5.11(a), (c)). In shorter gate length devices, this E_V maximum shifts deeper into the substrate to 8–10 nm for $L_G = 22$ nm. The corresponding hole density for this gate length (thick lines) of course also shows a maximum at the same position, causing the high OFF-state leakage.

The same figures for the IFQW pFET however show the full effect of the heterostructure confinement (Fig. 5.11(b), (d)). The valence band offset ΔE_V between the $Si_{0.50}Ge_{0.50}$ QW channel and the underlying Si substrate causes a sharp drop in hole density when crossing this hetero-interface. As such, only the SiGe QW channel has a noteworthy hole density, while the hole density in the Si substrate is well below 10^{10} cm^{-3}. Considering the continuity of the hole-quasi-Fermi level across the Si/SiGe interface (in equilibrium), the hole density in the Si is suppressed by about a factor $\approx 10^6$, using Eq. (5.1) ($\Delta E_V = 370$ meV, [155]). This factor corre-

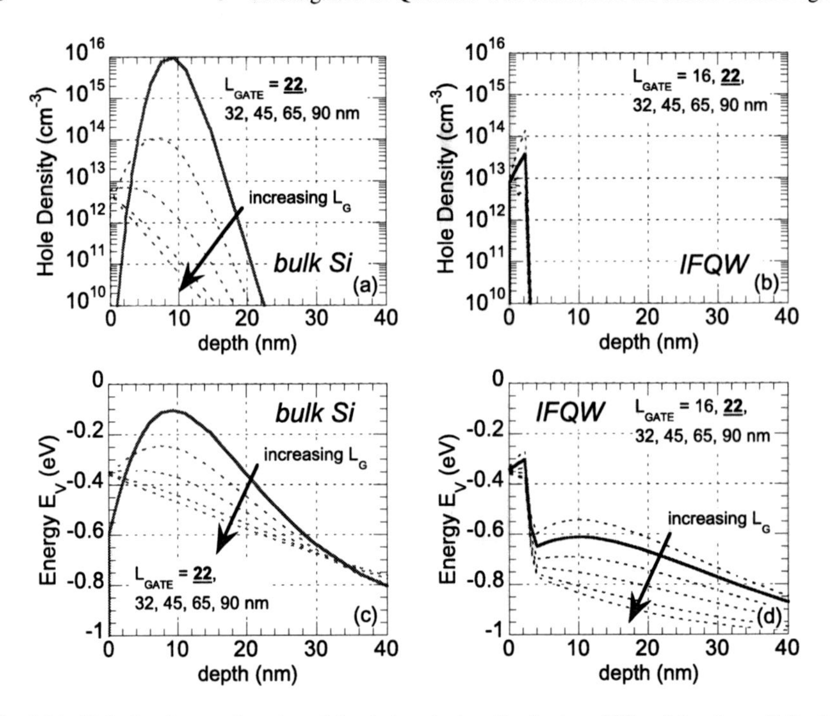

Fig. 5.11 Hole density as a function of depth (*vertical cutline* in the middle of the channel) for the bulk Si pFET and the IFQW pFET. Notice the effect of the valence band offset ΔE_V in the IFQW, drastically reducing the hole density in the Si substrate

sponds with the offset observed in the hole density in Fig. 5.11(b), although the factor is changed somewhat due to the applied bias conditions (non-equilibrium). In other words, holes are effectively confined to the SiGe layer by means of the valence band offset ΔE_V present at the SiGe/Si interface (bottom of the QW channel). As such, the ΔE_V is the main contributor to the strongly reduced hole density below the QW. Simultaneously, the absence of the implanted source/drain junctions below the QW avoids the undesired 2D proximity effect mentioned for the QW pFET [60].

A final set of figures to explain this heterostructure confinement is given in Fig. 5.12. Following the same methodology as with the bulk Si pFET (Fig. 5.4), 3D plots of the IFQW pFET (in OFF-state) were made with L_G ranging from 90 nm to 16 nm. The z-axis (pointing downwards) shows the valence band energy E_V as a function of the x and y coordinates, relative to the Fermi energy level E_F at the source. In this manner, a 3-dimensional surface of the energy barrier is generated that charge carriers (i.e. holes) have to overcome when flowing from source to drain. Notice the raised source/drain configuration (B-doped $Si_{0.75}Ge_{0.25}$), the Quantum Well channel (undoped $Si_{0.50}Ge_{0.50}$) and the low n-type doped Si substrate, having a valence band offset ΔE_V of about 370 meV to the QW channel.

For long channels, source and drain are separated by an energy barrier of 300 to 400 meV (green on the z-axis color scale), as was the case in the Si pFET plots (I_{OFF} will be low as few holes will be able to overcome this energy barrier). For

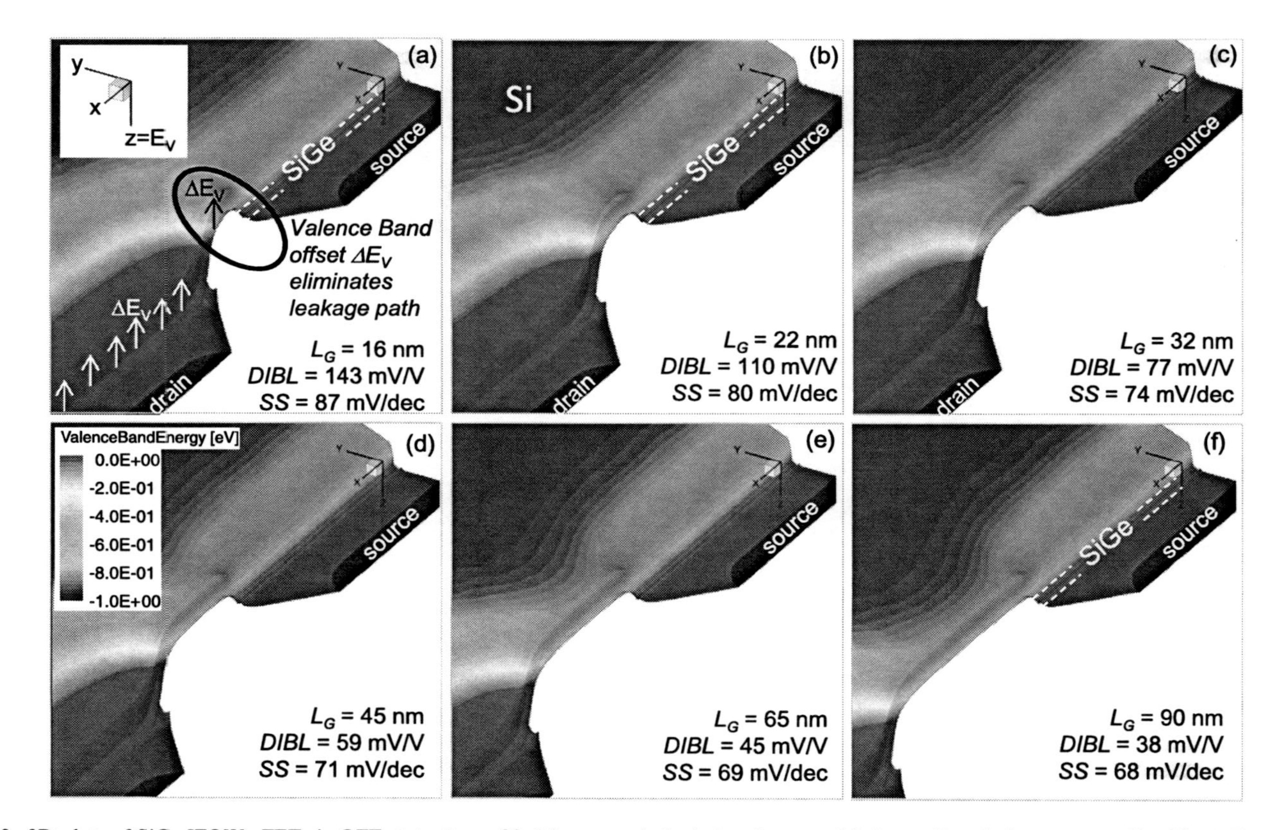

Fig. 5.12 3D-plots of SiGe IFQW pFETs in OFF-state, $L_G = 90$–16 nm. z-axis (pointing downwards) shows E_V relative to source E_F. The ΔE_V serves as a dam, isolating source and drain from each other

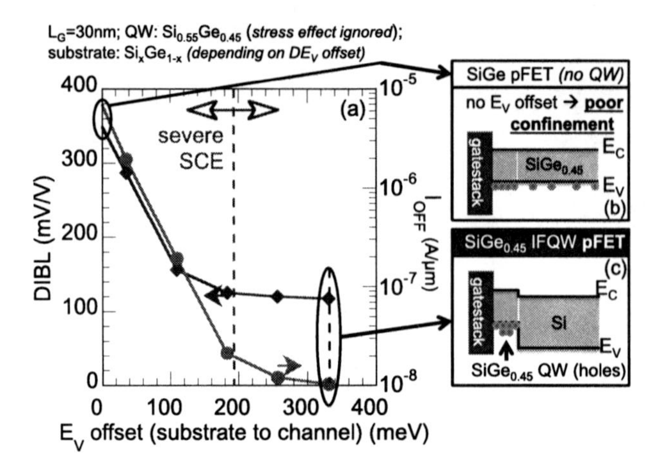

Fig. 5.13 (**a**) Simulated *DIBL* and I_{OFF} for the $L_G = 30$ nm $Si_{0.55}Ge_{0.45}$ IFQW pFET, as a function of ΔE_V between the QW channel and the substrate below. No further reduction in short channel control is observed pas $\Delta E_V \approx 200$ meV. (**b**), (**c**) Schematic of the band alignment for the situation with $\Delta E_V = 0$ meV and with $\Delta E_V = 330$ meV

shorter L_G, a saddle point appeared in the 3D surface, giving rise to the sub-surface leakage path (e.g. $L_G = 22$ nm, Fig. 5.4(a)). In contrast the energy barrier of 300–400 meV is maintained even at $L_G = 16$ nm for the IFQW pFET (Fig. 5.12(b)). The valence band offset ΔE_V creates a 370 meV high *energy wall* which the holes would have to overcome, giving rise to rather low *DIBL* and *SS* values, considering the small gate length of only 16 nm.

5.3.2.4 Heterostructure Confinement: Valence Band Offset ΔE_V

The preceding discussion has qualitatively explained the positive effect of heterostructure confinement on transistor scalability. This discussion however, also raises an important question on how large a band offset is really needed to have good short channel control in these transistors. An answer to this question is important when selecting the materials for the heterostructure. Attempting to answer this question, TCAD simulations were carried out on a SiGe IFQW pFET, varying valence band offset ΔE_V from 0 meV ($Si_{0.55}Ge_{0.45}$ on a $Si_{0.55}Ge_{0.45}$ substrate—IF pFET with raised source/drain but without a QW channel or band offset) to 330 meV ($Si_{0.55}Ge_{0.45}$ on a Si substrate—the IFQW pFET). These structures and material stacks may seem a bit strange at first. However, they are required to fully isolate the effect of the ΔE_V on transistor characteristics, keeping all other parameters constant as much as possible: channel material, EOT, source/drain structure. For the same reason, any effect of stress and strain on mobility was not included in these simulations.

As illustrated in Fig. 5.13(a), the structure without the band offset ($\Delta E_V = 0$ meV) suffers from large *DIBL*, and high I_{OFF}. These short channel effects gradually reduce until $\Delta E_V = 200$ meV. This value appears to yield a sufficient barrier to confine the charges to the SiGe QW channel: increasing ΔE_V beyond this point has little or no effect on *DIBL* and I_{OFF}.

5.3.3 Conclusions

In this section, the concept of heterostructure confinement was discussed in detail. A new transistor architecture, the Implant-Free Quantum Well transistor was introduced, using heterostructure confinement as a means to obtain better short channel control. Short channel effects are successfully suppressed by the band offsets in the heterostructure layer stack, as charge carriers are confined to a thin channel layer. More specifically, the combination of a Quantum Well channel and Implant-Free, epitaxially grown source/drain areas was shown to be essential in efficiently eliminating high OFF-state leakage currents at scaled gate lengths.

5.4 High Electron Mobility Transistors: an Alternative Approach

In the previous section, the Implant-Free Quantum Well FET was introduced, focussing on its implementation in the Si/SiGe material system. In the final section of this chapter a different approach is taken, introducing the Implant-Free Quantum Well FET starting from the High Electron Mobility Transistor (HEMT) on which its original design is based [84, 102].

Traditional HEMTs are optimized for high-frequency applications or fiber-optic front end systems and have a relatively large footprint. Particularly, the source/drain contacts are formed using lithography resulting in poor device pitch. In contrast, self-aligned processes are used in classical VLSI technologies, yielding minimal source/drain to channel separation. Addressing a second issue with classical HEMT technology, various studies [79, 130] have shown that reducing the distance between the gate and the Quantum Well channel is key to allow scaling below 100 nm gates. A third issue with classical HEMT technology is the absence of a dielectric layer between the gate electrode and the semiconductor materials underneath: a Schottky contact provides isolation between the gate electrode and the barrier layer.

5.4.1 HEMT with Interrupted Delta-Doping Layers

In a classical HEMT structure [84, 102], a delta doping layer is present between the gate electrode and the QW channel. Therefore, the influence of interrupting the HEMT δ-doping layer under the gate for a $L_G = 10$ nm device is studied. Interrupting this δ-doping would allow the gate to be closer to the channel, facilitating *EOT* reduction.

The structure used in this study is based on [102]. However, the conclusions should apply to a wide range of possible material combinations. On a GaAs substrate, an InGaAs quantum well with a thickness of 7 nm is modeled. The gate is assumed to be a Schottky contact ($\Phi_B = 0.8$ eV), and is separated from the QW by an AlGaAs spacer layer of 10 nm, which contains a continuous δ-doping layer. Laterally, the gate is isolated with a thin insulating layer (not shown). Figure 5.14(a) shows this first structure. Note that the gate has a classical T-shape and that lithography-defined source/drain contacts are directly contacting the QW channel.

Two alternative structures are analyzed here (Fig. 5.14(b) and (c)). Figure 5.14(b) presents the HEMT from Fig. 5.14(a), where the δ-doping layer has been interrupted over a length Ł$_\delta$. This allows the gate-to-channel distance h_{sp} to be reduced from 10 nm to 7 and 4 nm, leading to the structure in Fig. 5.14(c) (labeled E in Fig. 5.15(a).

Figure 5.15(a) shows the I_D–V_G curves for these structures, as simulated with Sentaurus Device [127]. The reference HEMT (−A) has a V_T of −0.19 V. Removing the δ-doping layer under the gate results in a V_T increase of 350 mV (Fig. 5.14, structure labeled B in the I_D–V_G plot). This can be explained by considering that an interruption of the δ-doping reduces the charge in the QW channel at a fixed V_G value. As a result, a less negative V_G will be required to effectively deplete the channel of electrons (and reach OFF-state). If needed, this can be mitigated by changing the δ-doping density [52, 53], as is done for structure C in the I_D–V_G plot. However this is not the focus of this study. Notice that the interruption of the δ-doping only results in a V_T shift: both structures (A and B) have very similar subthreshold behavior.

The advantage of an interruption in the δ-doping layer is that it allows further scaling of the spacing between the gate an the QW channel (h_{sp}). The effect of this is also visible in Fig. 5.15(a) when observing the I_D–V_G curves (labeled D and E) for the structure in Fig. 5.14(c) (resp. h_{sp} is reduced to 7 and 4 nm from 10 nm). This point becomes even more obvious when looking at Fig. 5.15(b), where *DIBL* and *SS* were plotted for all four structures. The large, positive effect of reducing h_{sp} is consistent with experimental results concerning a $L_G = 50$ nm HEMT, [79].

5.4.2 *Implant-Free Quantum Well FET*

The removal of the δ-doping layer in the HEMT structure solves an important issue with the scalability of these devices towards L_G = 10 nm. However, further improvements are clearly required, which are summarized in Fig. 5.14(d)–(f). First, the introduction of a gate dielectric would drastically reduce the gate leakage. Secondly, gate-first processing would allow self-aligned source/drains to be epitaxially grown, immediately next to the gate, drastically reducing the devices' footprint. Also, the removal of the Schottky-contact as gate electrode removes the requirement for the S/D material to have a large band gap (and high Φ_B towards the gate metal). This enables using the same material as the channel for the source and drain. And finally,

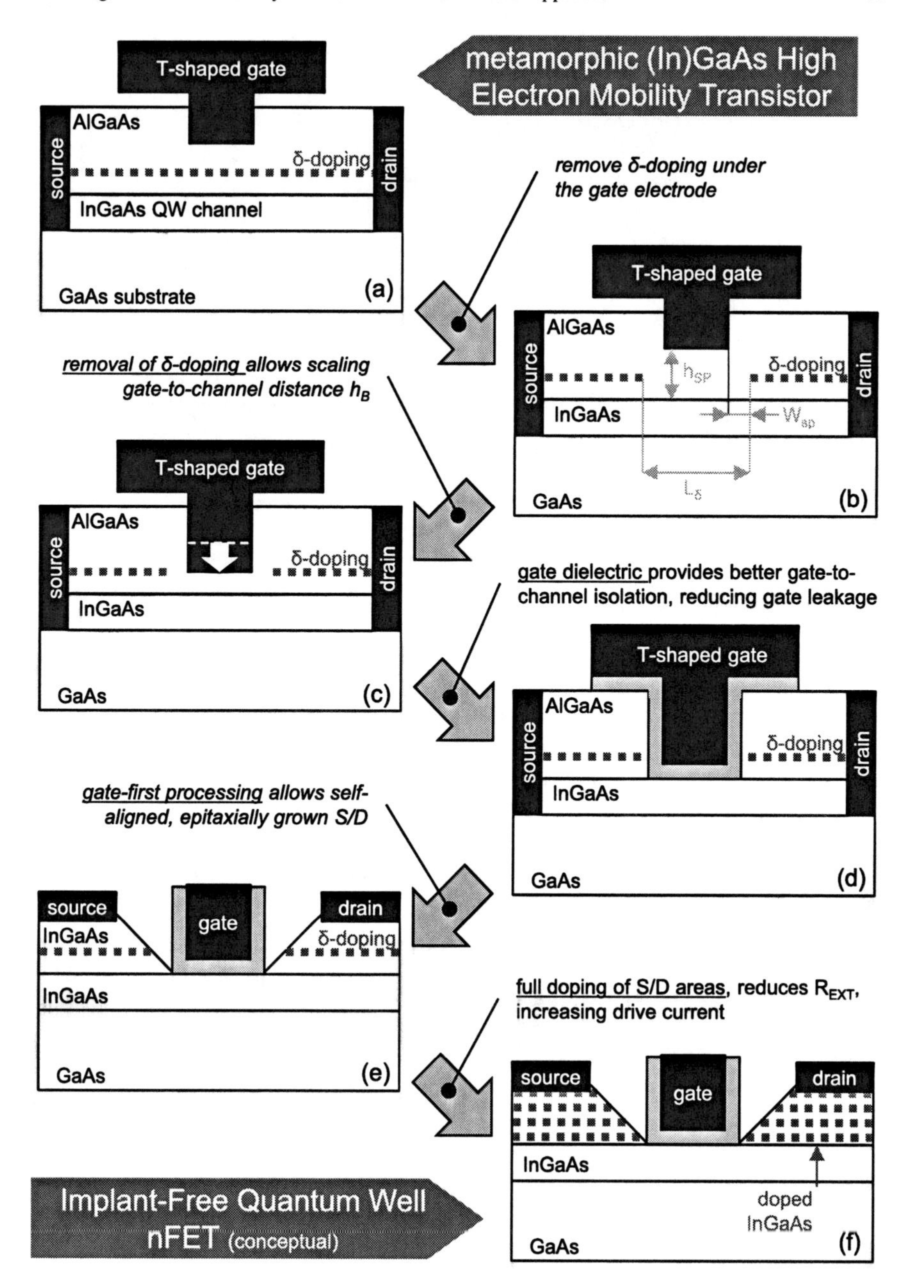

Fig. 5.14 Schematic overview of the (In)GaAs High Electron Mobility Transistor (HEMT) and various changes, ultimately leading to the Implant-Free Quantum Well Transistor (IFQW)

uniformly doping the epitaxially grown source/drains reduces the devices' external resistance, boosting drive current.

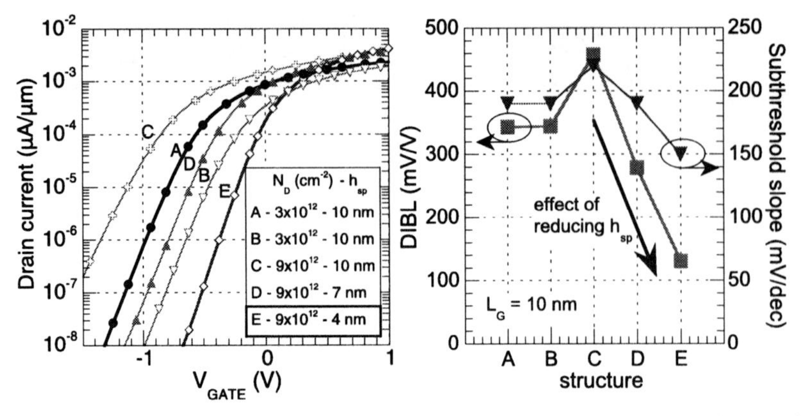

Fig. 5.15 Simulated I_D–V_G curves and *DIBL SS* for the reference HEMT (*A*) showing the effect of consecutively removing the δ-doping layer under the gate (*B*), increasing the δ-doping sheet density changing V_T (*C*) and reducing gate-to-channel spacing h_{sp} from 10 nm to 7 nm (*D*) and to 4 nm (*E*). Notice the absence of any effect on the short channel behavior when removing the delta doping (*A* to *B*) and the drastic improvement in short channel control when reducing h_{sp} (*C* to *D* to *E*)

As can be seen, the resulting structure (Fig. 5.14(f)) is a III-V nFET version of the Implant-Free Quantum Well pFET that was introduced using the Si/SiGe material system. The role of the valence band offset ΔE_V in the SiGe IFQW pFET is fulfilled here by the conduction band offset ΔE_C between the InGaAs channel and the GaAs substrate, both confining the carriers to the QW channel (holes resp. electrons). Finally, a similar comparison of different structures, starting from HEMT-like transistors was presented in [54] for the Si/SiGe/Ge material system. While the devices' external resistance was investigated in more detail, conclusions regarding short channel control are largely similar.

5.4.3 Conclusions

In this section, the Implant-Free Quantum Well nFET was introduced, starting from High Electron Mobility Transistor (HEMT). It has been shown that superior short channel control can be obtained by interrupting the δ-doping layer under the gate in the classical HEMT architecture. The removal of this δ-doping layer enables further EOT scaling for these devices. Finally, a gate-first, VLSI compatible fabrication scheme was proposed, featuring doped, self-aligned, epitaxially grown source/drains [54].

5.5 Operation of Heterostructure Transistors: Analytical Description

This section describes the basic electrical behavior of heterostructure transistors. Expressions for the flatband voltage, the depletion length and the threshold voltage

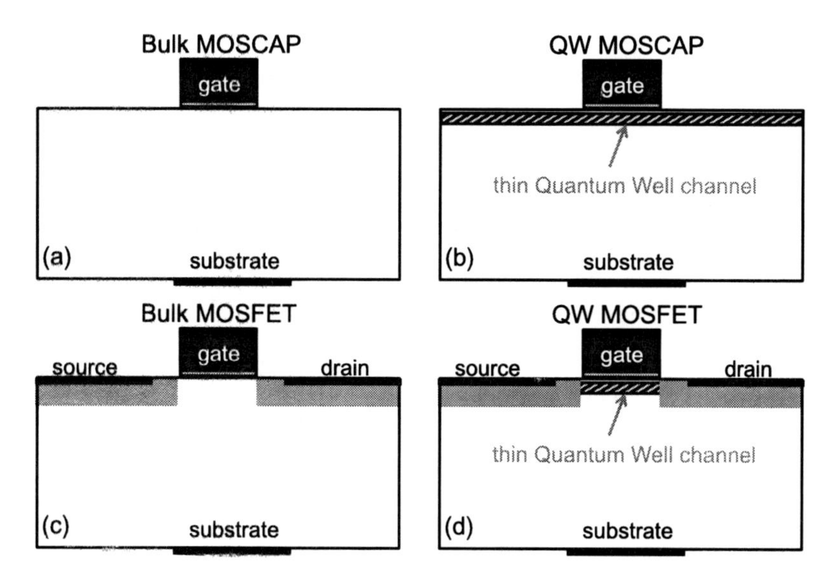

Fig. 5.16 Schematic of the different MOSCAP and MOSFET structures discussed in this section. Note the thin Quantum Well channel in the structures on the *right*

are derived on MOSCAP structures, MOSFET threshold voltage and body factor are derived on a 4-terminal QW MOSFET structure.

5.5.1 Transistor Structure

A schematic image of the four structures that will be discussed in this section is given in Fig. 5.16. On the left side, conventional bulk Si MOSCAP and MOSFET structures are shown. On the right side, the figure contains heterostructure MOSCAP and MOSFET structures, containing a thin Quantum Well below the gate dielectric. Note that although the SiGe material system can be used again as an example ($Si_{0.55}Ge_{0.45}$ QW channel overlying a Si substrate), the equations in this section are not restricted this specific application.

5.5.2 Approximations and Assumptions

The analytical description given in this section uses the approximations and assumptions that are usually needed to obtain concise equations for V_{FB}, V_{TH}, etc. [138]. Building on these, one additional approximation is added: The Quantum Well channel is considered to be extremely thin. As such, it can only contain an inversion or accumulation charge. (Note that both are also considered to be extremely thin in the

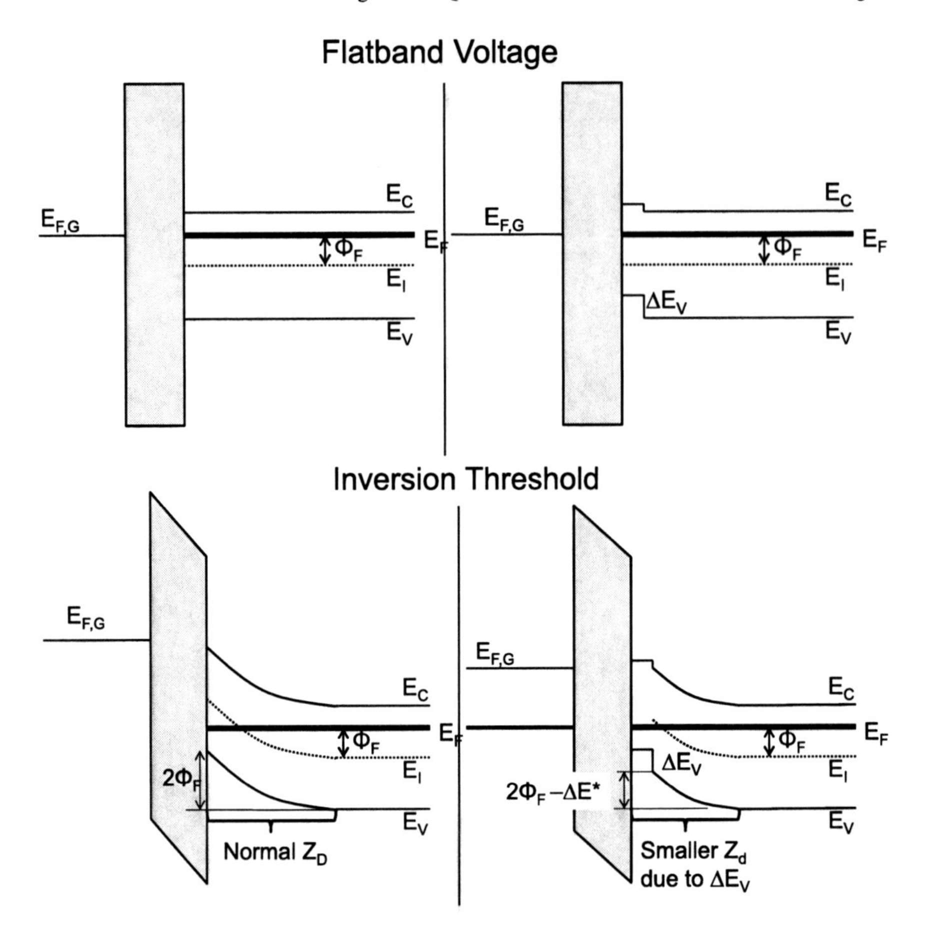

Fig. 5.17 Band diagram of the conventional bulk MOSFET (*left*) and the SiGe QW MOSFET (*right*) in flat band condition (*top*) and at inversion threshold (*bottom*). Notice the reduced substrate band bending required to reach inversion threshold in the QW MOSFET. For simplicity, V_{FB} was taken to be 0 V and any difference in DOS between QW and substrate was ignored in this figure

classical derivation [138].) This also means that the Quantum Well cannot contain a depletion charge. Any electrostatic potential drop inside the QW is neglected in the following calculations.

5.5.3 MOS Capacitor

Energy-band diagrams for ideal n-type MOS capacitors under different bias conditions are shown in Fig. 5.17. For simplicity, the flatband voltage is taken to be zero in the plots.

5.5.3.1 Flatband Voltage: V_{FB}

By definition, V_{FB} is the gate voltage for which the substrate bands are completely flat. In a conventional MOSCAP, this voltage is given by:

$$V_{FB} = \Phi_{MS} - \frac{Q_{OX} - Q_{SS}}{C_{OX}} \tag{5.2}$$

In this equation, Φ_{MS} is the workfunction difference between gate and semiconductor, Q_{OX} is the fixed equivalent oxide charge, Q_{SS} contains charged interface states and C_{OX} is the dielectric capacitance. This expression (Eq. (5.2)) is also applicable to QW-based MOSCAP, assuming Φ_{MS} is taken to be the workfunction difference between the gate and the substrate (not between the gate and the QW channel).

5.5.3.2 Depletion

When a gate voltage is applied, such that the majority carriers are pushed away from the gate dielectric, a depletion region occurs in the semiconductor. In a conventional MOSCAP, the potential drop over the gate oxide is given by:

$$\Delta\Psi_{OX} = -\frac{Q_{OX} - Q_{SS}}{C_{OX}} + \frac{q(N_A - N_D)Z_D}{C_{OX}} \tag{5.3}$$

Here, N_A and N_D are the acceptor and donor concentrations in the substrate, while Z_D is the depletion depth. The potential drop over the semiconductor (substrate) is given by:

$$\Delta\Psi_S = \frac{q(N_A - N_D)Z_D^2}{2\epsilon_S} \tag{5.4}$$

where ϵ_S is the relative permittivity of the semiconductor. These equations are both also still valid for the heterostructure MOSCAP, since the QW itself was assumed to not contain any depletion charge. Note that all parameters of Eq. (5.4) relate to the substrate material in QW MOSFETs.

5.5.3.3 Inversion Threshold

When the gate voltage is increased further, a significant concentration of minority carriers are attracted to the gate dielectric. The gate voltage at which their local concentration equals that of the majority carriers in the substrate is the threshold voltage V_{TH}. In regular MOSCAP structures, this condition occurs when the potential drop over the semiconductor reaches a fixed value of:

$$\Delta\Psi_S = 2\phi_F \quad \text{(conventional MOSCAP)} \tag{5.5}$$

Here, ϕ_F has the usual definition of $kT/q \ln([N_A - N_D]/n_i)$. For a heterostructure MOSCAP, the required band bending is different. A difference in valence/conduction band energy (between QW and substrate) and in density of states for the two materials yields a modified expression. Following the classical definition, the onset of inversion is defined as the gate voltage for which the local concentration of minority carriers in the channel exceeds that of the majority carriers in the substrate (e.g. for nFET):

$$p_{substrate} = n_{QW} \tag{5.6}$$

or:

$$N_{V,substrate} \exp\left(\frac{E_F - E_V}{kT/q}\right) = N_{C,QW} \exp\left(\frac{E_C - E_F}{kT/q}\right) \tag{5.7}$$

where $N_{V,substrate}$ and $N_{C,QW}$ are the density of states in the substrate valence band and the QW conduction band respectively. As such, it can be shown that the potential drop over the substrate can be expressed as:

$$\Delta\Psi_S = 2\phi_F + \Delta E^* \quad \text{(QW MOSCAP)} \tag{5.8}$$

with:

$$\Delta E^* = E_{C,substrate} - E_{C,QW} + kT/q \ln(E_{C,substrate}/N_{V,substrate}) \tag{5.9}$$

In this equation, the last term is negligible when both materials have comparable density of states. This is the case in the SiGe material system. In typical heterostructure MOSCAPs, the potential drop over the semiconductor is smaller.

Combining Eqs. (5.9) and (5.4), one can find an expression for the depth of the depletion layer Z_D:

$$Z_D = \sqrt{\frac{2\epsilon_S(2\phi_F + \Delta E^*)}{q(N_A - N_D)}} \tag{5.10}$$

leading to the following equation for the threshold voltage V_{TH} in the heterostructure MOSCAP:

$$V_{TH} = V_{FB} + 2\phi_F + \Delta E^* \pm \frac{\sqrt{2\epsilon_S(2q\phi_F + \Delta E^*)(N_A - N_D)}}{C_{OX}} \tag{5.11}$$

(+ for p-type semiconductor, − for n-type semiconductor). Notice that this equation is similar to that for the regular MOSCAP except that $2\phi_F$ has been replaced by $2\phi_F + \Delta E^*$. Following the above definition, this quantity is often smaller in absolute value (using typical material systems used for QW).

5.5.3.4 MOSCAP Operation Regions

Equation (5.11) can be used to determine the different regions of operation for the MOSCAP (accumulation, depletion and inversion). These have been plotted for

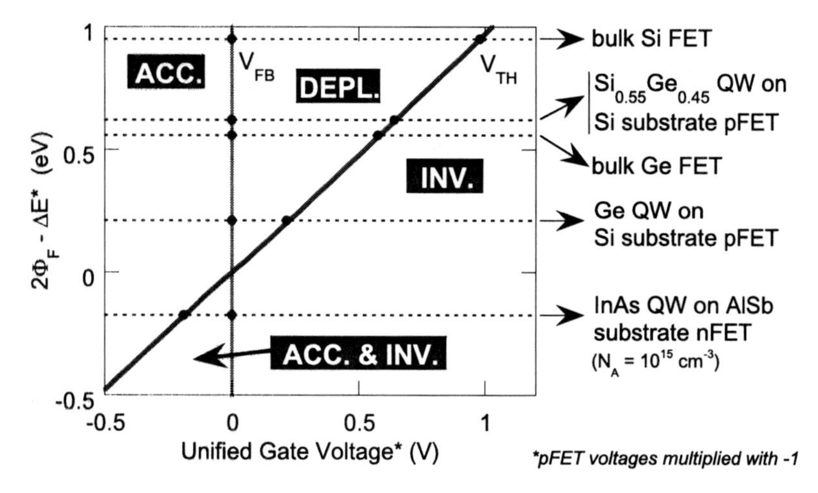

Fig. 5.18 Graphical representation of the different regions of operation in conventional bulk MOSFETs and QW MOSFETs. The quantity on the y-axis is the required substrate band bending and is set by the different materials, determining the different regions. Note that in the bottom case (InAs QW), effectively no depletion region exists

some MOSCAP structures in Fig. 5.18 as a function of (unified) gate voltage (V_{FB} was taken to be 0 V, for simplicity). The required potential drop over the substrate to reach inversion (Eq. (5.9)) is plotted on the y-axis. This way, each MOSCAP structure's operation regions fall onto a horizontal line, going from accumulation to depletion to inversion. Note that V_{FB} and V_{TH} are closer together the bulk Ge, compared to the bulk Si (due to the smaller bandgap). The use of most heterostructures are also observed to bring V_{FB} and V_{TH} closer together. Substrate doping density was taken to be 10^{18} cm^{-3} in this plot, unless indicated otherwise.

A peculiar feature of Eq. (5.11) is that under certain conditions, no solution exists. This occurs when ΔE^* becomes larger than $2\phi_F$, making the argument of the square root negative. In those MOSCAPs, no depletion operation region exists, effectively merging the inversion and accumulation regions. Whatever gate voltage is applied, no band bending is observed. An example material combination where this is expected to occur is an InAs QW on top of a AlSb substrate (p-type doped).

5.5.4 MOS Field Effect Transistor

5.5.4.1 Inversion Threshold V_{TH}

Following a similar formalism as before, it can be shown that at the onset of inversion, the total band bending at the source of a QW MOSFET can be expressed as:

$$\Delta\Psi_S = 2\phi_F + \Delta E^* - V_{BS} \tag{5.12}$$

where V_{BS} is the potential difference between the source and the bulk terminal. The depth of the depletion region is then equal to:

$$Z_D = \sqrt{\frac{2\epsilon_S(2\phi_F + \Delta E^* - V_{BS})}{q(N_A - N_D)}} \tag{5.13}$$

The MOSFET threshold voltage is then given by:

$$V_{TH} = V_{FB} + 2\phi_F + \Delta E^* \pm \frac{\sqrt{2\epsilon_S q(2\phi_F + \Delta E^* - V_{BS})(N_A - N_D)}}{C_{OX}} \tag{5.14}$$

5.5.4.2 Body Factor

In a MOSFET the Body Factor is defined as the ratio between the bulk-to-channel capacitance and the gate-to-channel capacitance. It is given by:

$$\frac{C_{BC}}{C_{GC}} = \frac{\epsilon_S\sqrt{q(N_A - N_D)}}{C_{OX}\sqrt{2\epsilon_S(2\phi_F - V_{BS})}} \tag{5.15}$$

For QW MOSFETs, the body factor is equal to (from Eq. (5.10)):

$$\frac{C_{BC}}{C_{GC}} = \frac{\epsilon_S\sqrt{q(N_A - N_D)}}{C_{OX}\sqrt{2\epsilon_S(2\phi_F + \Delta E^* - V_{BS})}} \tag{5.16}$$

Comparing Eqs. (5.15) and (5.16), it can be seen that the body factor is affected by the presence of the QW: the smaller depletion width Z_D in QW MOSFETs (compared to their conventional counterpart) effectively increases the body factor through the denominator of Eq. (5.16).

5.5.5 Conclusions

In this section, the operation of heterostructure transistors was discussed. Analytical expressions were derived for the threshold voltage and the body factor. In typical QW MOSFETs, the relevant band offset between the substrate and the QW channel was found to decrease the amount of band bending required to reach inversion in the QW channel. Since this also decreases the depth of the depletion layer Z_D, the channel-to-bulk capacitance is also increased. In turn, this leads to a larger body factor in QW MOSFETs (i.e. the threshold voltage depends more strongly on the Bulk potential V_B.

5.6 Conclusions

In this chapter, scaling issues and short channel effects in the bulk MOSFET were analyzed using TCAD simulations. Firstly, the drain-to-bulk junction leakage in bulk germanium MOSFET technology was investigated. A trade-off was found between good short channel control and low junction leakage for a 65 nm germanium pFET technology. Extension leakage current densities below the ITRS spec of 100 nA/μm could only be obtained for supply voltages of 0.7 V or lower (keeping the reference halo implant). As smaller gate lengths would require heavier halo implants, it seems unlikely that a planar bulk Ge MOSFET technology would be well suited for future technology nodes ($L_G < 20$ nm). Secondly, the effect of imposing fixed source/drain junctions and a maximum halo implant dose were investigated on a scaling silicon pFET. Investigating the cause for the severely degraded I_D–V_G characteristics for these transistors, revealed a sub-surface leakage path from source to drain, causing high *DIBL* and *SS* values. As a consequence, when increasing the halo dose in bulk Si technology is no longer an option, alternative transistor architectures will also be required.

In the second part of this chapter, a class of transistors was introduced, where charge carriers are confined to a Quantum Well (QW) by means of heterostructure confinement. The Implant-Free Quantum Well FET was presented and its scaling performance analyzed for gate lengths down to 16 nm using the Si/SiGe material system as an example. Even at this small gate length, simulated *DIBL* and *SS* remained quite low at $\approx$150 mV/V and 90 mV/dec respectively, markedly lower than for the bulk Si pFET. Finally, the role of the valence band offset between the $Si_{0.50}Ge_{0.50}$ channel and the Si substrate underneath was investigated in the SiGe IFQW pFET. Using TCAD simulations, a band offset of 200 meV between the QW channel layer and the substrate material was shown to suffice in controlling control short channel effects. An InGaAs/GaAs IFQW nFET was also introduced with the High Electron Mobility Transistor as starting point, showing good short channel control at a gate length of $L_G = 10$ nm.

Finally, analytical expressions were derived for the depth of the depletion layer, the threshold voltage and the body factor in QW MOSFETs.

Chapter 6
Implant-Free Quantum Well FETs: Experimental Investigation

In this chapter, following the design considerations of Implant-Free Quantum Well FETs presented in the previous chapter, such transistors are fabricated and electrically characterized.

6.1 Introduction

As scaling progresses, higher mobility channel materials such as SiGe and Ge are being considered to boost transistor performance [157]. Simultaneously, alternative structures such as multi-gate FETs or SOI-based devices [21] are being developed to maintain electrostatic gate control. In the previous chapter, a third class of transistors was introduced, using a heterostructure to confine the charge to a Quantum Well. TCAD simulations predicted that the combination of this QW channel with epitaxially grown source/drains can be very successful in suppressing short channel effects.

This chapter describes the process development to fabricate the $Si_{0.55}Ge_{0.45}$ Implant-Free Quantum Well pFET. A first section will discuss such IFQW pFETs with raised source/drains. These *first-generation* SiGe IFQW pFETs are electrically analyzed, focussing on the scalability and short channel behavior of such a technology. The second section of this chapter will focus on enhancing the performance of *first-generation* IFQW pFETs. Specifically, the integration of source/drain stressors into the fabrication process will be investigated, aiming to boost channel mobility and reduce the external resistance of such transistors. In a third section, *second-generation* $Si_{0.55}Ge_{0.45}$ Implant-Free Quantum Well pFET are then presented. These improved IFQW pFETs incorporate an additional embedded $Si_{0.75}Ge_{0.25}$ source/drain, delivering uniaxial compressive strain to the transistor channel. These devices are benchmarked against a state-of-the-art planar bulk Si pFET technology. The IFQW pFETs' performance is also compared to that of strained Silicon-On-Insulator nFETs.

In the fourth section of this chapter, the matching performance and V_T-tuning capabilities of an IFQW pFET technology are investigated. Varying well doping

G. Hellings, K. De Meyer, *High Mobility and Quantum Well Transistors*,
Springer Series in Advanced Microelectronics 42, DOI 10.1007/978-94-007-6340-1_6,

across a chip and applying bulk biasing to certain circuit blocks are known techniques to reduce static power consumption. The feasibility of such power management techniques (on IFQW pFETs) are explored. In the fifth and final section, the SiGe Quantum Well itself is analyzed in detail. Being a core feature of SiGe IFQW pFET technology, the physical properties of this layer are investigated, focussing on Si-Ge interdiffusion processes which could potentially impact the devices' electrical performance.

Note that the classification of the IFQW pFETs discussed in this chapter into *first-* and *second-generation* devices is purely arbitrary. It merely serves the purpose of orienting the reader and clearly distinguishing between two specific embodiments of SiGe IFQW pFETs discussed in this book.

Finally, this chapter focusses on key experimental results obtained on a SiGe IFQW pFET technology. Additional information can be found in literature: SiGe IFQW pFET technology in [11, 40, 54, 57–62, 64, 82, 101, 160], III/V IFQW nFET technology in [8–10, 12, 13, 52, 53].

6.2 First-Generation SiGe Implant-Free Quantum Well pFET

In this section the fabrication of the first-generation $Si_{0.55}Ge_{0.45}$ Implant-Free Quantum Well pFET is discussed. In this first embodiment, the SiGe IFQW pFET is integrated on bulk Si wafers, featuring a $Si_{0.55}Ge_{0.45}$ Quantum Well channel and raised, in-situ B-doped $Si_{0.75}Ge_{0.25}$ source/drains. Following this discussion, the transistors' electrical behavior is analyzed and benchmarked against standard bulk Si pFET devices with the same gate stack.

6.2.1 Device Concept and Fabrication

Fabrication of the Implant-Free Quantum Well pFET starts from a bulk Si substrate (n-well 2×10^{17} cm^{-3}). Following Shallow Trench Isolation (STI), a 3 nm undoped, epitaxial $Si_{0.55}Ge_{0.45}$ layer is grown on this starting substrate (in an ASM Epsilon 3200 epi system) forming the QW channel. This QW is subsequently topped with an in-situ grown Si cap for surface passivation [66, 100]. To avoid surface roughening the epitaxial layers have been grown at low temperature ($<$500 °C) using SiH_4 and GeH_4 as precursor gases. H_2 is used as carrier, and a low growth pressure ($<$40 Torr) is used to avoid loading effects [91].

After gate-stack processing (SiO_2/high-κ/Metal Gate/poly Si), a thin offset spacer is deposited on the gate sidewalls. The thickness of this offset spacer is about 3 nm (measured at the bottom of the gate) and the spacer etch has been observed to also remove the Si cap and SiGe QW in the source/drain areas in these IFQW pFETs. Faceted, in-situ B-doped ($\sim 10^{20}$ cm^{-3}) $Si_{0.75}Ge_{0.25}$ source/drains are epitaxially grown to a thickness of 20–30 nm and partially silicided. In the devices

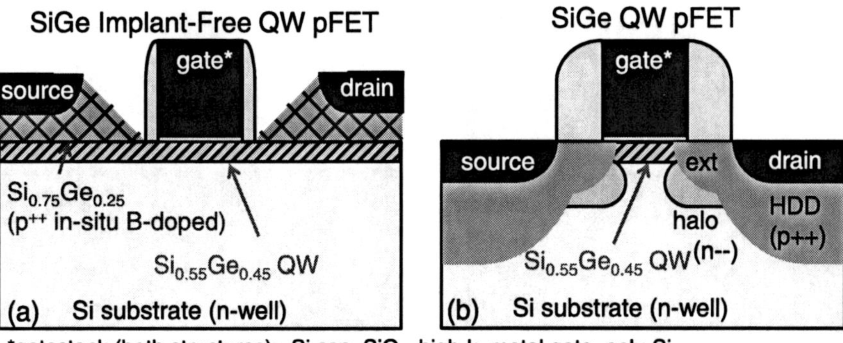

Fig. 6.1 Schematic view of the SiGe Implant-Free Quantum Well pFET (**a**) and the SiGe Quantum Well pFET (**b**)

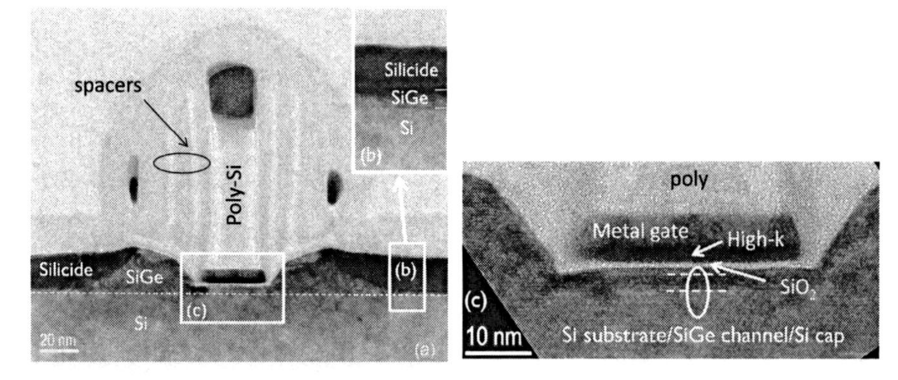

Fig. 6.2 (**a**) TEM image of the $Si_{0.55}Ge_{0.45}$ IFQW pFET, featuring a defect-free SiGe channel, thin offset spacer and raised, facetted, in-situ B-doped source/drain. The physical gate length is 30 nm. After silicidation, approximately 6–8 nm of B-doped SiGe remains under the silicide (close-up in (**b**)). (**c**) High-resolution TEM close-up of the $L_G = 30$ nm IFQW pFET. HAADF-STEM and ellipsometry measurements (not shown) indicate a SiGe QW channel thickness of 3 nm (sample preparation FIB lift-out, Tecnai F30 operating at 300 kV)

presented in this section, a mild (950 °C) spike anneal was included. However, detailed physical analysis confirmed a negligible impact of this thermal step on the Ge concentration in the QW channel (this will be discussed in detail in Sect. 6.6). Note that no dopants (halos, extensions etc.) were implanted after the deposition of the $Si_{0.55}Ge_{0.45}$ QW channel. The reported well doping values should be considered estimates, since actual doping concentrations vary as a function of depth (due to the distribution as a function of depth resulting from the well ion implant steps).

This first-generation IFQW pFET is shown schematically in Fig. 6.1(a). A TEM image of the $L_G = 30$ nm SiGe IFQW pFET is shown in Fig. 6.2. Notice the thin offset spacer and the facetted epitaxially grown source-drains. The Scanning Spreading Resistance Microscopy (SSRM) image in Fig. 6.3 visualizes the low-resistive, in-

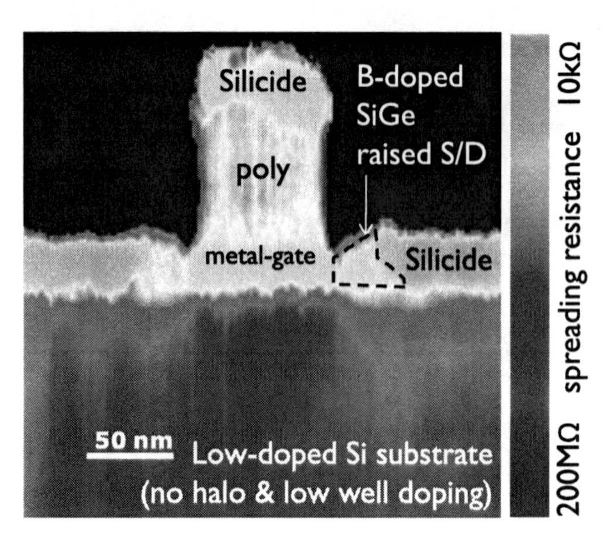

Fig. 6.3 SSRM-image (Scanning Spreading Resistance Microscopy) of the $Si_{0.55}Ge_{0.45}$ IFQW pFET. Notice the highly-resistive Si substrate, silicided S/D and the B-doped S/D. The thin QW channel itself could not be resolved. High Vacuum SSRM, full diamond tip, cleaved sample [41]

situ B-doped $Si_{0.75}Ge_{0.25}$ S/D (partially silicided). Additionally, the low doping in the Si substrate is confirmed by SSRM measurements with a local spreading resistance of $\sim$100 MΩ.

In this section, IFQW pFETs are compared against conventional SiGe QW channel pFETs (with implanted halo and S/D doping and corresponding, high-temperature spike anneal, Fig. 6.1(b)) and Si control pFETs, all with the same gate stack (SiO_2/high-κ/Metal Gate/poly Si).

6.2.2 Electrical Results and Discussion

In Fig. 6.4(a), the I_D–V_G characteristics of a typical $L_G = 30$ nm IFQW pFET are plotted. This device, having a total width of 1 μm, showed a saturation drive current I_{ON} of 582 μA/μm ($V_G = V_D = -1$ V) and an OFF-state drain current I_{OFF} of 100 nA/μm ($V_G = 0$ V, $V_D = -1$ V). I_D–V_D curves are plotted in Fig. 6.4(b) for $V_G - V_T$ ranging from 0 to −1.2 V in steps of 200 mV. The threshold voltage V_T shows an almost flat behavior as a function of L_G in Fig. 6.5(a) around a value of −0.25 V, on target for high-performance logic applications [75]. For the Si control pFET (same gatestack), V_T is about 400 mV lower around −0.6 V. The on-target V_T for the IFQW pFET eliminates the need for a V_T-adjusting cap layer which is needed to adjust the V_T in regular Si pFETs. The *DIBL* plot (Fig. 6.5(b)) illustrates the excellent short channel control in the IFQW pFET, with typical $L_G = 30$ nm devices having a *DIBL* value of 126 mV/V. *DIBL* is markedly higher in the Si control pFET and in the SiGe QW pFET. The improved short channel behavior is further illustrated by a rather low subthreshold swing *SS* of 80 mV/dec ($V_D = -1$ V, $L_G = 30$ nm) in Fig. 6.4(a).

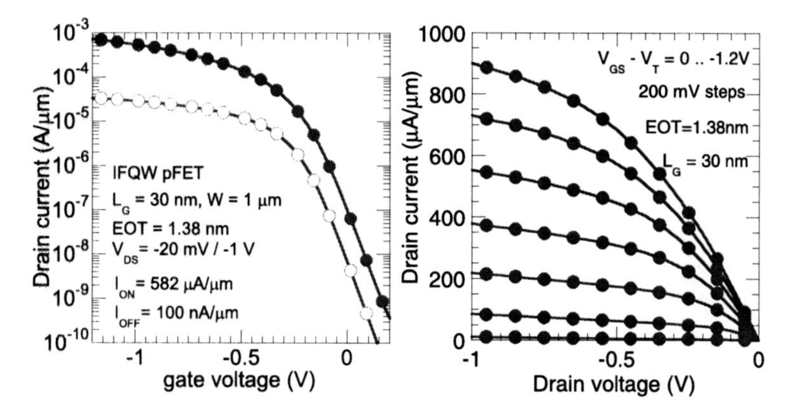

Fig. 6.4 (**a**) Measured linear and saturation I_D–V_G curves for the SiGe Implant-Free Quantum Well pFET ($L_G = 30$ nm, channel: $Si_{0.55}Ge_{0.45}$). (**b**) Measured I_D–V_D curves for the SiGe IFQW pFET, for different overdrive values ($V_G - V_T$ ranging from 0 to −1.2 V)

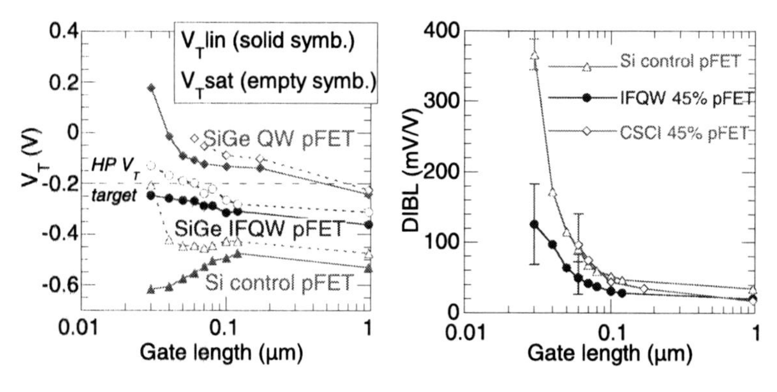

Fig. 6.5 (**a**) Linear and saturation threshold voltage (V_T and $V_{T,sat}$) as a function of gate length for the SiGe IFQW pFET, the SiGe QW pFET and the Si control pFET. (**b**) *DIBL* as a function of gate length for the same devices

I_{ON}–I_{OFF} plots at a fixed gate voltage and at a fixed gate overdrive are presented in Fig. 6.6. The SiGe IFQW shows a ~50 % higher I_{ON} than the bulk-Si control and the SiGe QW pFET devices with the same gatestack (at a fixed I_{OFF} of 100 nA/μm, conforming to the ITRS specifications for high performance logic). The observed I_{ON}–I_{OFF} performance and electrostatic control resembles that achieved on extremely-thin, fully depleted SOI pFETs ($L_G = 25$ nm, [21]), while the bulk-Si based flow reduces the wafer cost for the IFQW. Note also that the drain-to-bulk leakage is similar to that of the Si reference device with the same geometry, several orders of magnitude below the HP logic I_{OFF} requirements (Fig. 6.7(a)). As expected this leakage can be lowered by a factor 10× by lowering the well doping in the IFQW pFET to 2×10^{16} cm^{-3}. Plotting gate delay (CV/I) as a function of the I_{ON}/I_{OFF} ratio in Fig. 6.7(b) (using the benchmark from [19]) further illus-

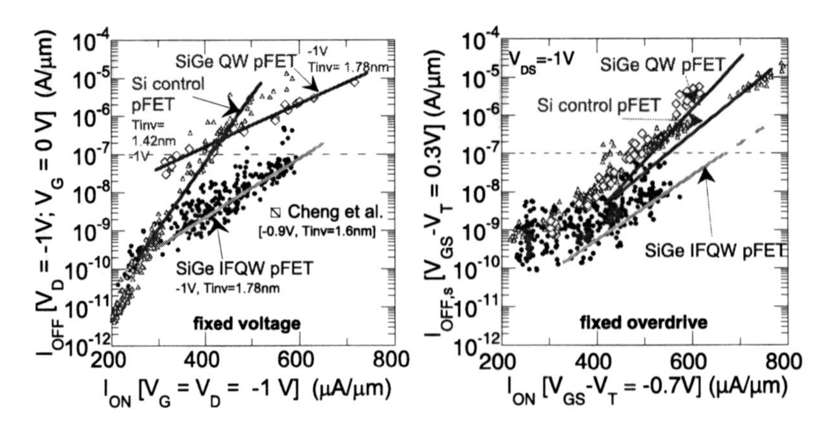

Fig. 6.6 I_{ON}–I_{OFF} plot for the $Si_{0.55}Ge_{0.45}$ IFQW pFET, a $Si_{0.55}Ge_{0.45}$ QW pFET and a Si control pFET (all with the same gatestack) at fixed voltage (*left*) and at fixed gate overdrive (*right*). Notice the low I_{OFF} for the IFQW devices. An UTSOI reference with raised source/drain was added for comparison [21]

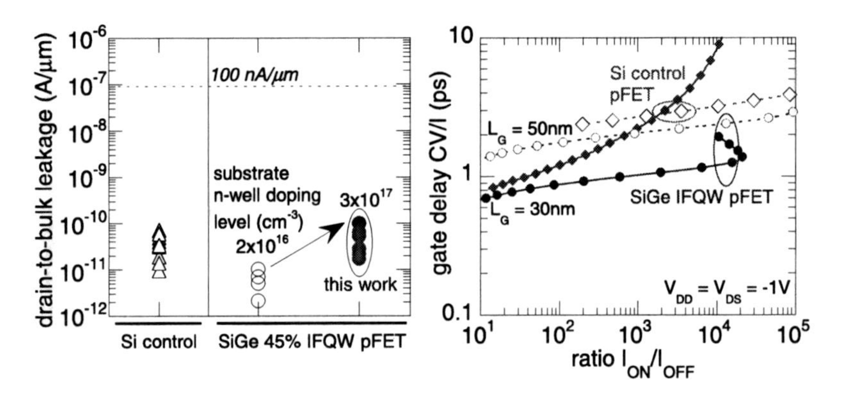

Fig. 6.7 (*left*) Drain-to-bulk junction leakage (normalized to pFET width) as a function of substrate doping level for the $Si_{0.55}Ge_{0.45}$ IFQW pFET. (*right*) Gate delay (CV/I) as a function of I_{ON}/I_{OFF} ratio (sliding V_{DD} window, see Ref. [19] for the IFQW pFET and Si ref. devices. Note the lower gate delay for the IFQW devices, especially for high I_{ON}/I_{OFF} ratios and smaller L_G

trates the electrostatic control and high drive current in the IFQW pFETs, leading to improved gate delays, especially for high I_{ON}/I_{OFF} ratios and short L_G.

High-frequency C-V analysis was also performed on these devices, yielding a Capacitance Equivalent Thickness of $CET = 1.78$ nm for the IFQW pFET and the SiGe QW pFET and a lower $CET = 1.42$ nm for the Si control pFET (Fig. 6.8). This slightly higher CET value can be attributed to the presence of the Si cap layer, which is present on the SiGe QW channel in the first two transistors, possibly in combination with reduced interlayer oxide scavenging due to the low-temperature processing [119]: a high-temperature spike (>1000 °C) dopant activation anneal is included in the fabrication process.

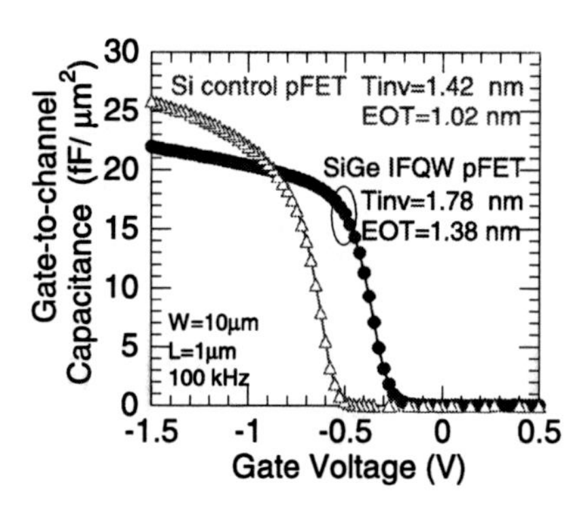

Fig. 6.8 Capacitance-Voltage measurement of the $Si_{0.55}Ge_{0.45}$ IFQW pFET and the Si ref. pFET. *CET* and *EOT* were extracted at a fixed gate overdrive ($V_G = V_T - 0.7$ V), and are slightly higher for the IFQW pFET because of the Si cap layer and the absence of interlayer (IL) scavenging due to the limited thermal budget

6.2.3 Conclusions

In this section, the first-generation $Si_{0.55}Ge_{0.45}$ Implant-Free Quantum Well pFET was presented. Fabricated devices with gate lengths down to $L_G = 30$ nm were electrically analyzed and compared to conventional bulk silicon pFETs with an identical gate stack. Excellent short channel control was observed with *DIBL* and *SS* values of 126 mV/V and 80 mV/dec respectively for $L_G = 30$ nm IFQW pFETs. A 50 % higher saturation drive current was obtained, comparing to Si control pFETs. As such, the combination of a Quantum Well channel (heterostructure confinement) and Implant-Free, epitaxially grown source/drain areas have been experimentally shown to greatly suppress short channel effects.

6.3 Enhancing Performance in SiGe IFQW pFETs

The first-generation SiGe IFQW pFETs discussed in the previous section have successfully shown that good short channel behavior can be obtained using this transistor structure. However, while the saturation drive current $I_{ON} = 582$ μA/μm was shown to be markedly higher than that of an unstrained Si reference pFET with identical L_G and gate stack, it is still well below that of a state-of-the-art 32 nm technology (e.g. $I_{ON} = 1200$ μA/μm, Intel [153]). Further improving the IFQW pFET is therefore very important. This section therefore focusses on enhancing the performance of the first-generation SiGe IFQW pFETs presented in the previous section.

In order to improve transistor performance, reducing the series resistance has been identified as an important challenge in many alternative FET architectures, often requiring innovative solutions [21]. Secondly, source/drain stress engineering has proven to be a highly manufacturable technique for improving transistor performance [48], essential in today's competitive technologies. As such, this section investigates the *integration of source/drain stressors* into the IFQW pFET technology, aiming to boost channel mobility and to reduce the external resistance. In a later stage, other performance-enhancing techniques should be investigated (reduction of the *EOT* to boost gate-to-channel capacitance, or increasing the channel Ge content).

In this section, incremental improvements from first- to second-generation SiGe IFQW pFETs will be discussed. The detailed electrical analysis of the second-generation SiGe IFQW pFETs will be discussed in the next section.

6.3.1 Experimental Details

Three different versions of the SiGe IFQW pFET transistor will be fabricated. All start from bulk-Si substrates with Shallow Trench isolation (n-well 2×10^{17} cm^{-3}) on which a 3 nm thick undoped epitaxial $Si_{0.55}Ge_{0.45}$ Quantum Well is grown and capped with a thin, in-situ Si cap. Following gatestack fabrication a 3 nm offset spacer is formed against the gate sidewalls. Then,

- to obtain *structure A*, facetted B-doped $Si_{0.75}Ge_{0.25}$ source/drains are epitaxially grown to a thickness of 30 nm. This $Si_{0.75}Ge_{0.25}$ is then partially silicided, leaving approximately 6–8 nm of $Si_{0.75}Ge_{0.25}$. A TEM image of this structure was already shown in Fig. 6.2(a), as it is the first-generation SiGe IFQW pFET, as discussed in the previous section.
- to obtain *structure B*, facetted B-doped $Si_{0.75}Ge_{0.25}$ source/drains are epitaxially grown to a thickness of 50 nm. This $Si_{0.75}Ge_{0.25}$ is then partially silicided, leaving a thicker $Si_{0.75}Ge_{0.25}$ layer of about 26–28 nm.
- to obtain *structure C*, facetted B-doped $Si_{0.75}Ge_{0.25}$ source/drains are epitaxially grown to a thickness of 30 nm. After this epitaxial deposition, a second spacer is formed against the gate sidewall. The raised $Si_{0.75}Ge_{0.25}$ and the Si substrate are then recessed ($\approx$50 nm) and replaced by a second, thick $Si_{0.75}Ge_{0.25}$ epi-layer to form embedded source/drain regions. Also this material is then partially silicided. A TEM-image of this structure is shown in Fig. 6.9.

Conventional back-end modules conclude the fabrication process for all structures.

6.3.2 Source/Drain Stressors and TCAD Modeling

6.3.2.1 Strain Models and Calibration

The stress and strain in structure C with the embedded source/drain was simulated using a commercial TCAD simulator [128]. The resulting longitudinal stress profile σ_{XX} is plotted in Fig. 6.10(a), clearly showing the compressive stress in the channel, beneficial for channel hole mobility. To check the calibration of the simulations, Nano Beam Diffraction measurements were carried out [42] on a cutline going from source to drain at a depth of 25 nm below the gate. The measured longitudinal (ϵ_{XX})

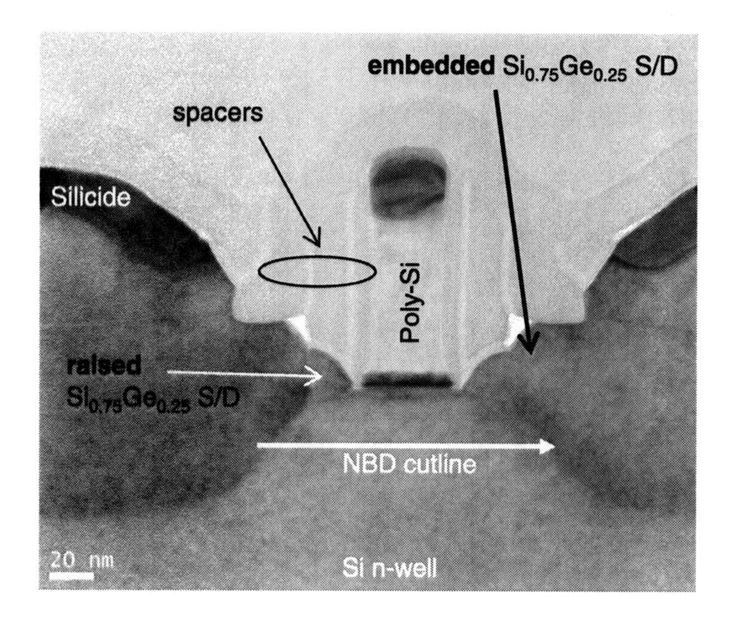

Fig. 6.9 TEM image of the SiGe Implant-Free Quantum Well pFET with embedded $Si_{0.75}Ge_{0.25}$ source/drains (structure C)

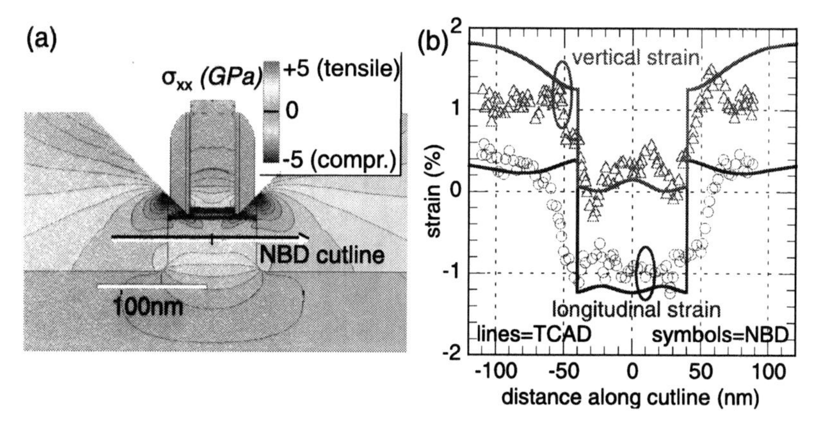

Fig. 6.10 (**a**) TCAD-simulated longitudinal stress profile σ_{XX} for IFQW pFET in Fig. 6.9 (structure C). (**b**) Longitudinal and vertical strain as measured by Nano-Beam Diffraction (NBD) and as simulated across a cutline 25 nm below the gate oxide for this device. The reported NBD strain is normalized with respect to the (relaxed) Si substrate

and vertical (ϵ_{ZZ}) strain values were then compared to the simulations. As shown in Fig. 6.10(b), a good quantitative agreement was obtained for both strain components, revealing a compressive longitudinal strain ϵ_{XX} of about 1 %, in combination with a near-zero vertical strain ϵ_{ZZ} along this buried cutline (below the gate). Note that these simulations were carried out in 2D, making them valid for wide pFETs.

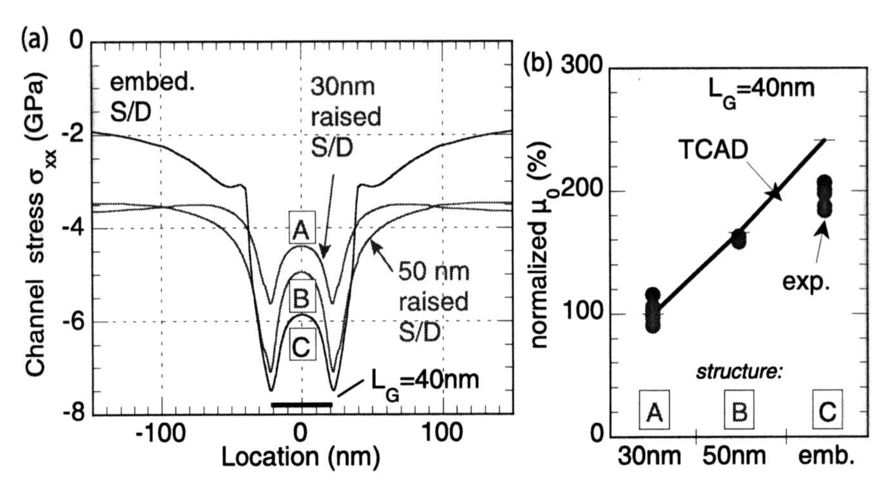

Fig. 6.11 (**a**) Simulated stress profile for the three IFQW structures. (**b**) Expected channel hole mobility improvement for each IFQW pFET, relative to structure A

6.3.2.2 Hole Mobility Increase

Here, we will attempt to give an estimate of the mobility enhancement that can be expected for a particular source/drain structure. As such, the three orthogonal stress components were extracted along a horizontal cutline through the $Si_{0.55}Ge_{0.45}$ Quantum Well channel. These orthogonal components are then averaged across the length of the channel, after which they are combined to calculate the change in mobility using standard piezoconductance:

$$\vec{\delta\mu} = \vec{\Pi} \cdot \vec{\sigma_{av}} \quad \text{with } \sigma_{av_i} = \frac{1}{L_G} \int_{L_G} \sigma_{ii} \cdot dx \quad \text{for } i = X, Y, Z \tag{6.1}$$

$\vec{\Pi}$ is the piezoconductance tensor for $\langle 011 \rangle/(100)$ $Si_{0.55}Ge_{0.45}$ pFET channels, using a linear interpolation of values reported by Smith et al. for holes in Si and Ge [133].

The simulated longitudinal stress σ_{XX} in the $Si_{0.55}Ge_{0.45}$ quantum well for each of the simulated structures is plotted in Fig. 6.11(a). As expected, a compressive longitudinal channel stress is observed in all structures (the average σ_{XX} across the channel is 4.69, 5.45 and 6.3 GPa for structures A, B and C respectively). In comparison, the biaxial stress in a thin $Si_{0.55}Ge_{0.45}$ layer on a relaxed Si substrate is about 3.7 GPa (lattice mismatch 1.8 %). As such, the highest compressive stress values are obtained in structure C, with the embedded $Si_{0.75}Ge_{0.25}$ source/drains. Using the aforementioned methodology, the expected mobility increase was calculated for each of the simulated structures (Fig. 6.11(b)), using structure A as a reference. As expected, structure B offers the smaller mobility increase of 66 %, while the embedded S/D in structure C is expected to boost the mobility by 141 %, relative to structure A.

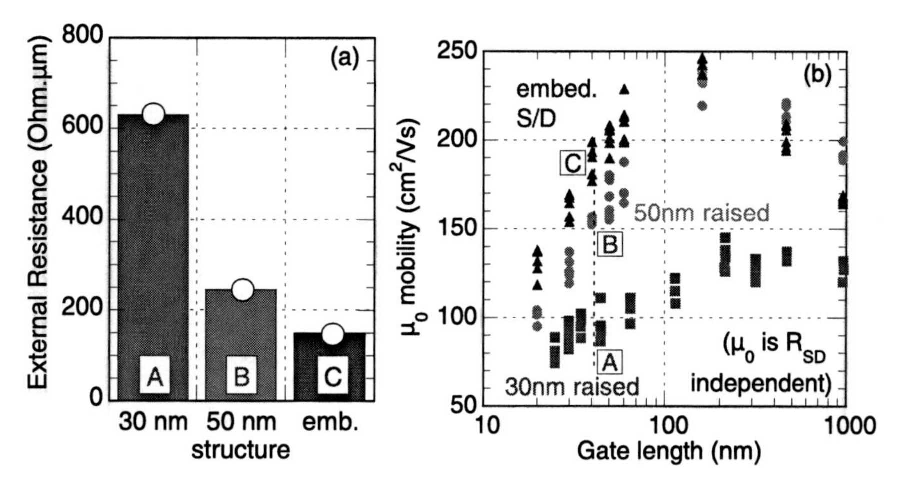

Fig. 6.12 (**a**) Measured external resistance R_{EXT} for the IFQW pFETs and (**b**) measured intrinsic mobility μ_0 as a function of L_G. Note the reduced R_{EXT} and higher μ_0 for structure C, with embedded $Si_{0.75}Ge_{0.25}$ source/drains

6.3.3 Electrical Results and Discussion

6.3.3.1 External Resistance and Channel Mobility

When comparing devices with a different S/D architecture, it is important to separate the effect of channel mobility and external resistance. Both were extracted on our devices. Figure 6.12(a) shows R_{EXT}, where a clear reduction is visible when moving from structure A to B to C. As the offset spacer thickness—an important contributor to R_{EXT} [101]—is identical in all structures, the main reason for this reduction is thought to be reduced current crowding: near-full silicidation of the raised S/D in A results in a small useful area for the SiGe/silicide interface, a problem which is avoided by increasing raised S/D thickness (B) or introducing embedded S/D (C). Investigating the effect of channel mobility, the R_{EXT}-independent μ_0 was extracted using the Y-method introduced by Fleury et al. [45]. The μ_0-boost from the S/D stressors is clearly visible in Fig. 6.12(b), showing a maximal hole mobility of 240 cm^2/Vs for structures B and C at $L_G = 160$ nm. The mobility boost in short-channel pFETs ($L_G = 40$ nm) matches very well with the predicted μ_0 boost from TCAD, as shown in Fig. 6.11, although the observed relative increase is slightly lower than simulated for structure C (+∼90 %). This offset is likely due to the approximation (linear interpolation) made in Sect. 6.3.2.2.

6.3.3.2 Transistor Performance

The aforementioned R_{EXT} reduction, combined with the mobility increase results in increased I_{ON}–I_{OFF} performance (Fig. 6.13) with saturation drive current I_{ON} of 600, 800 and 1000 µA/µm for the three structures at ITRS HP-target

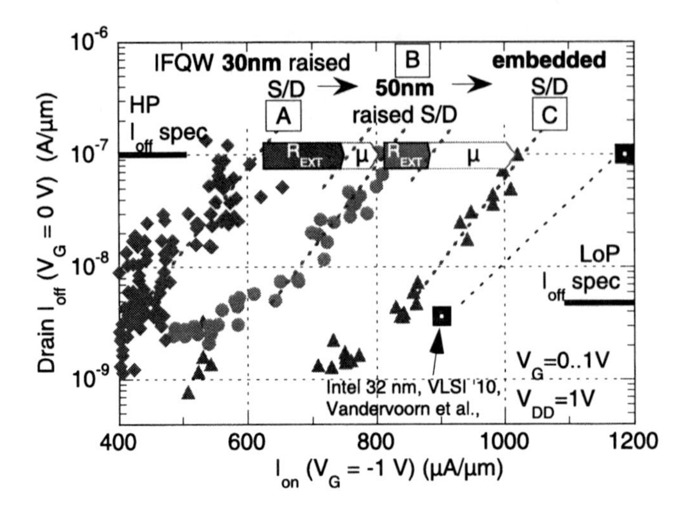

Fig. 6.13 I_{ON}–I_{OFF} plot for the IFQW structures. The performance boost from A to B is mainly due to R_{EXT} reduction, while the embedded source/drain in C is most successful in boosting mobility

$I_{OFF} = 100$ nA/μm. Using a first-order model, the main part (∼80 %) of the I_{ON} improvement in B (compared to A) can be attributed to reduced R_{EXT}. The improvement from B to C, on the other hand, is mainly due to the mobility boost (only 35 % because of the R_{EXT} reduction). Obviously, the co-optimization of both components is required to achieve a high drive current. Finally, note that the carrier confinement in the IFQW transistor architecture still results in good short-channel control for structure C, despite the embedded source/drain: DIBL values between 100–120 mV/V are observed ($L_G = 35$ nm), for all IFQW pFET structures discussed.

Finally, while the observed saturation drive current $I_{ON} = 1000$ μA/μm of structure C is still about 20 % lower than that of a current state-of-the-art planar strained-Si technology [153], it should be noted that (among other optimizations) the IFQW pFET performance can probably be improved by reducing *EOT*: in the presented IFQW technology, the *EOT* value of ∼1.38 nm is still quite relaxed, as compared to the state-of-the-art gate stack (*EOT* = 0.95 nm or ∼30 % lower in [153]). With these considerations in mind, the SiGe-IFQW with embedded S/D can be considered to be competitive with current state-of-the-art planar strained-Si technologies.

6.3.4 Conclusions

In this section, raised and embedded $Si_{0.75}Ge_{0.25}$ source/drain stressors were integrated into the Implant-Free Quantum Well pFET architecture. The effect on channel mobility and external resistance was investigated using calibrated TCAD simulations. Compared to the first-generation IFQW pFETs, a raised source/drain with a thickness of 50 nm was found to boost short-channel mobility by 58 %, while the embedded S/D approximately doubles it, as confirmed by experimental data. Through simultaneous R_{EXT} reduction, the SiGe-IFQW with embedded S/D is shown to be competitive with current state-of-the-art planar strained-Si technologies [153], exhibiting excellent short-channel control.

6.4 Second-Generation Strained SiGe IFQW pFETs

In this section, the $Si_{0.55}Ge_{0.45}$ Implant-Free Quantum Well pFETs with embedded source/drain will be discussed in detail. Discussed in the previous section as *structure C* (Fig. 6.9), it incorporates an additional embedded $Si_{0.75}Ge_{0.25}$ source/drain delivering uniaxial compressive strain to the transistor channel.

6.4.1 Device Fabrication

The key fabrication steps of the $Si_{0.55}Ge_{0.45}$ Implant-Free Quantum Well pFET with embedded source drain were already discussed in Sect. 6.3.1. Compared to the first-generation IFQW devices [57], the raised SiGe and Si substrate are recessed and replaced by a thick, B-doped $Si_{0.75}Ge_{0.25}$ epi-layer to form source and drain. A good interface quality between this new epitaxial layer and the underlying substrate is indispensable to obtain fully-strained $Si_{0.75}Ge_{0.25}$ source/drains: on the TEM image in Fig. 6.9, no dislocation defects were observed at this interface. An SSRM image for this IFQW pFET is shown in Fig. 6.14 and shows limited boron out-diffusion from the epitaxially grown source/drain towards the Si substrate underneath, due to the limited thermal budget in this processing flow.

6.4.2 Electrical Results and Discussion

As shown in Fig. 6.15(a), the I_D–V_G characteristics of the $L_G = 35$ nm IFQW pFET demonstrate the good short channel control (*DIBL* = 110 mV/V), in combination with a high saturation drive current $I_{ON} = 1015$ μA/μm at $V_G = V_D = -1$ V. As mentioned before, this high I_{ON} is reached, despite the fact that the *EOT* in these pFETs is still quite relaxed, compared to state-of-the-art gate stacks [1, 119, 157]. I_{ON}–I_{OFF} performance at $V_{DD} = -1$ V of this second-generation IFQW pFETs was reported in Fig. 6.13 and shows that the ITRS I_{OFF} specifications for Low Operating Power (LoP: $I_{OFF} = 5$ nA/μm, [75]) technologies can also be met for a drive current $I_{ON} \approx 850$ μA/μm.

At a slightly lower supply voltage of $V_{DD} = 0.9$ V, benchmarking of this IFQW technology shows that the pFET device has similar performance to strained SOI nFETs (Fig. 6.16(left)). This similarity is maintained, even when the I_{EFF} metric is used, instead of I_{ON}. This metric was introduced to predict invertor delay more

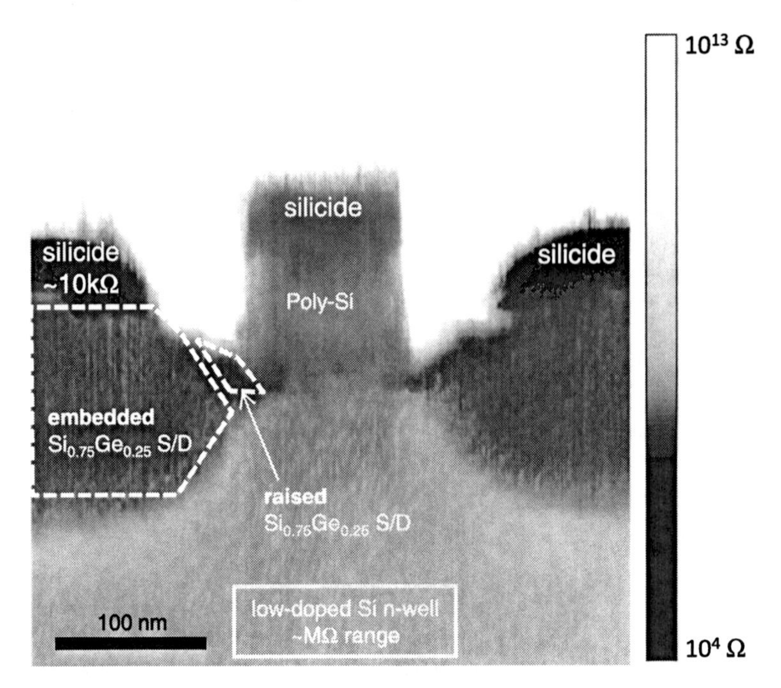

Fig. 6.14 SSRM-image (Scanning Spreading Resistance Microscopy) of the $Si_{0.55}Ge_{0.45}$ IFQW pFET with embedded $Si_{0.75}Ge_{0.25}$ source/drains. Notice the highly-resistive Si substrate and B-doped S/D. The thin QW channel itself could not be resolved, due to the proximity of the metal gate

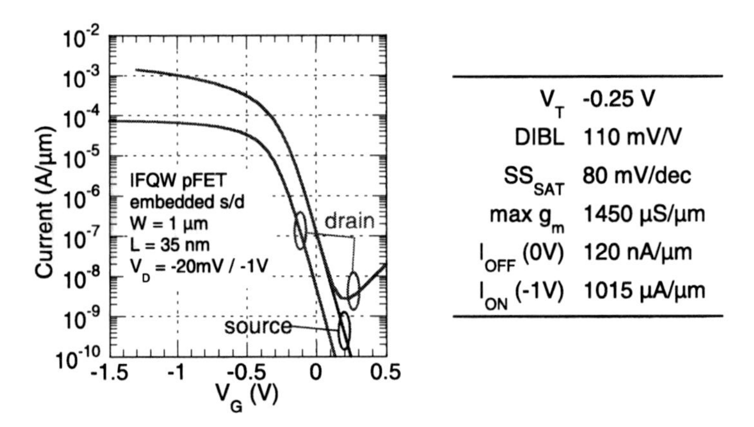

V_T	-0.25 V
DIBL	110 mV/V
SS_{SAT}	80 mV/dec
max g_m	1450 μS/μm
I_{OFF} (0V)	120 nA/μm
I_{ON} (-1V)	1015 μA/μm

Fig. 6.15 I_D–V_G curves of a typical $L_G = 35$ nm IFQW pFET with embedded source/drains in linear and saturation regimes. *The table* lists various relevant quantities extracted on this device

accurately, considering that both gate and drain voltage are significantly smaller than V_{DD} during most of the invertor switching process [104]. Compared to the more conventional I_{ON} metric, I_{EFF} also includes short channel control. The low *DIBL*

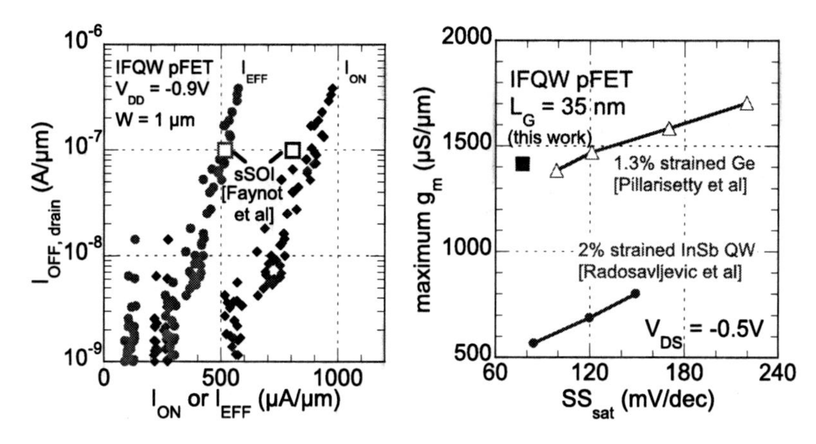

Fig. 6.16 (*left*) I_{ON} and I_{EFF} vs I_{OFF} performance at $V_{DD} = 0.9$ V, for the second-generation SiGe IFQW pFET and benchmarking with sSOI nFETs [43], showing matched performance. (*right*) Peak transconductance $g_{m,max}$ as a function of subthreshold slope in saturation of the second-generation SiGe IFQW pFET and other high-mobility channel FETs from literature ($V_{DD} = 0.5$ V, [115, 118])

values in our SiGe IFQW pFET devices, in combination with the strain-enhanced channel mobility yields similar I_{EFF} than that in strained SOI nFETs.

Finally, the performance of high-mobility channel FETs such as strained-Ge and III-V based devices are often benchmarked at an even lower $V_{DD} = 0.5$ V. At this low operating voltage, Fig. 6.16(right) shows that the intrinsic performance of the strained $Si_{0.55}Ge_{0.45}$ IFQW pFET outperforms the state-of-the art strained Ge pFETs [115], despite a serious long channel mobility penalty (160 cm^2/Vs in the IFQW pFETs versus 770 cm^2/Vs at $N_{INV} = 5 \times 10^{12}$ cm^{-2}).

6.4.3 Conclusions

In this section, the *second-generation* strained $Si_{0.55}Ge_{0.45}$ IFQW pFET with embedded source/drain was presented. Excellent short channel control was maintained with *DIBL* values of 110 mV/V at $L_G = 35$ nm and a high saturation drive current I_{ON} of 1 mA/µm. Increased performance at lower operating voltage was demonstrated and comparison with strained-SOI technology nFETs showed competitive performance. Considering that there is still significant room for further improvement of the IFQW pFET, this comparison suggests that it should be considered a viable device option for upcoming technology nodes.

6.5 Matching Performance and V_T-Tuning in IFQW pFETs

This section will explore the matching performance and V_T-tuning capabilities of an IFQW pFET technology. In order to make circuits more power-efficient, designers

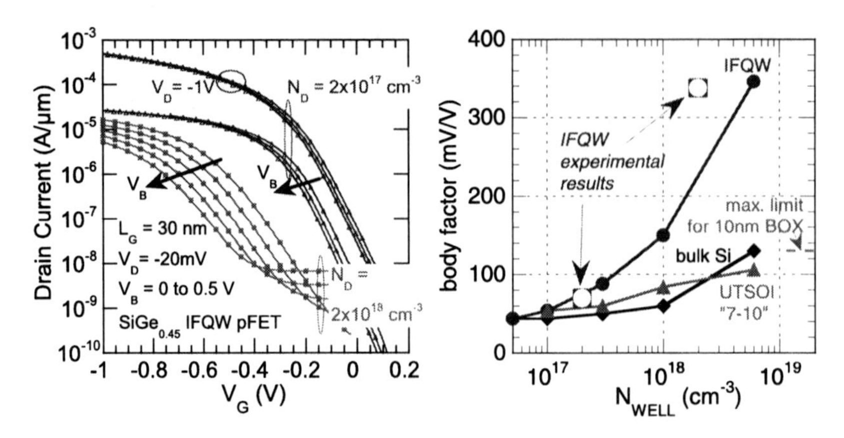

Fig. 6.17 (*left*) Experimental I_D–V_G curves for the $L_G = 30$ nm SiGe IFQW pFET, changing the bulk bias from 0 to 500 mV. The body factor is about 70 mV/V for the reference condition ($N_{WELL} = 2 \times 10^{17}$ cm^{-3}) and increases to 338 mV/V for a higher well doping ($N_{WELL} = 2 \times 10^{18}$ cm^{-3}). In the latter case, drain-to-bulk junction leakage currents are visible, due to the large junction area in our test devices (200 μm^2). (*right*) TCAD simulated body factor as a function of well doping concentration for the SiGe IFQW pFET, UTSOI (7 nm channel, 10 nm BOX) and bulk Si ($L_G = 30$ nm). Experimental results show reasonable agreement with the simulations. The reported body factor is the average V_T change, varying V_B from 0 to 0.5 V

need to be able to change the V_T of their transistors either during operation or at the time of fabrication. The former allows dynamic power management, whereby certain parts of an IC can be temporarily put in a state in which leakage is reduced. The latter allows a differentiation in transistors whereby a lower V_T can be tolerated for e.g. high-speed transistors in a core while a higher V_T reduces leakage in the rest of the IC. Such a technology option is often referred to as *multi-V_T technology*.

6.5.1 Body Bias Sensitivity

In classical bulk Si FETs, V_T can be changed during operation by applying a body bias. The body bias sensitivity of two $L_G = 30$ nm SiGe IFQW pFETs (first-generation with raised S/D) was measured experimentally in Fig. 6.17(left), containing I_D–V_G characteristics for different bulk bias conditions ($V_B = 0$ to 500 mV). The body factor is observed to be around 70 mV/V for the reference IFQW pFET with a Si n-well doping of $N_{WELL} = 2 \times 10^{17}$ cm^{-3}. It increases sharply to 338 mV/V for a higher well doping (2×10^{18} cm^{-3}). For this second (somewhat extreme) condition, the drain-to-bulk leakage is also observed to sharply increase. While this is due to the large drain junction area in our test devices (200 μm^2), such junction leakage currents will ultimately impose an upper limit on the amount of well doping that can be included in an IFQW technology.

The body factor was also simulated using TCAD for the SiGe IFQW pFET, a bulk Si technology and an ultra-thin Silicon-On-Insulator (UTSOI) technology,

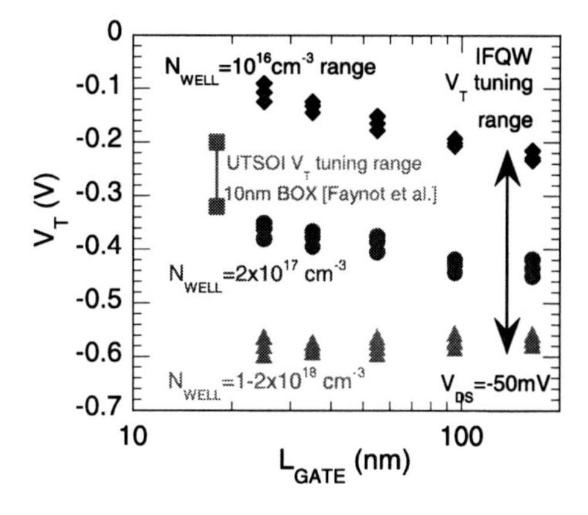

Fig. 6.18 Experimental V_T (L_G), showing the effect of a change in N_{WELL}. The IFQW shows a wide V_T-tuning range, which is maintained at short L_G. This range is smaller on UTSOI structures, as most of the effect is lost as a potential drop across the BOX

as a function of well doping in Fig. 6.17(right). For the UTSOI, a Si channel thickness of 7 nm was used, with a 10 nm buried oxide, after [43]. The well doping of the underlying Si substrate was varied. Notice that for the classical bulk pFET, the body factor increases with higher well doping. The same trend is observed in the IFQW pFET. As such, a high body factor can be obtained for the IFQW pFET if desired by increasing the well doping. Measurements from Fig. 6.17(left) were also included and show reasonable agreement with the simulations. For the UTSOI, the body factor is limited due to the buried oxide: its maximum value is determined by the ratio of the channel-to-substrate capacitance over the channel-to-gate capacitance ($C_{BC}/C_{GC} = 130$ mV/V for our example with an *EOT* of 1 nm). On the other hand, an advantage of having a buried oxide is that a very large V_B-tuning range can be applied without causing excessive leakage currents. As such, the wider V_B-tuning range can compensate for the lower body factor and ultimately may result in a larger dynamic V_T-tuning range in UTSOI-technology.

6.5.2 Multi-V_T Technology

Changing V_T at the time of fabrication can be done by modifying the well doping, as is typically the case in classical bulk Si FETs. As shown in Fig. 6.18, the IFQW architecture offers a very large V_T tuning range of 350 mV as a function of well doping. This sensitivity is maintained at short channels. In UTSOI, due to the presence of the buried oxide, a much smaller sensitivity is observed: changing the substrate doping from high n-type to high p-type has, in comparison, a rather small effect. The reason for this is again that most of the potential drop occurs over the buried oxide.

6.5.3 V_T Matching Performance

Variability of threshold voltage V_T is widely stated as one of the critical challenges for the future CMOS technology nodes [4]. Random dopants in the channel are

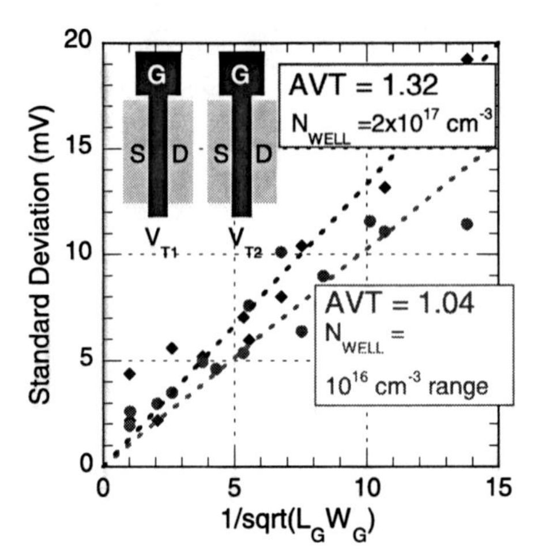

Fig. 6.19 Pelgrom plot for the first-generation SiGe IFQW pFET, showing improved V_T-matching, especially when further lowering well doping N_{WELL}

known to be a major source of the local V_T variability in the current bulk MOSFET technology. As a result, transistor architectures without high channel doping (well implant and halo doping) should have reduced local V_T variability.

Transistor mismatch properties have been experimentally characterized and shown to be proportional to the inverse square root of the active gate area [111]. This has led to the classical Pelgrom plot, showing $\sigma(\Delta V_T)$ as a function of $1/\sqrt{L_G W_G}$. The slope of the linear relationship (often referred to as *AVT*) is a measure for a technology's V_T matching performance. As shown in Fig. 6.19, the absence of halo doping and relatively low n-well doping in the IFQW pFET is clearly beneficial for the device's matching performance. While a typical value for *AVT* in an optimized bulk Si technology is close to 2.0 [83], a lower value of 1.32 was observed in the IFQW pFET technology. Further lowering the Si well doping from 2×10^{17} cm^{-3} to 1×10^{16} cm^{-3} reduces the *AVT* parameter to 1.04 (Fig. 6.19). The latter indicates that the active gate area required can be $4\times$ smaller in the IFQW technology, compared to bulk Si for similar V_T-matching. Note that improved V_T mismatch improvements have also been reported in undoped UTSOI devices [2] and finFETs [6], clearly indicating that a low channel doping is key to reduced V_T variability.

6.5.4 Conclusions

In this section, the matching performance and V_T-tuning capabilities of IFQW pFETs have been investigated. This technology has been shown to provide significant V_T-tuning capabilities. Firstly, during circuit operation a high V_T sensitivity to the bulk bias can be obtained for devices with increased well doping. Secondly, at the time of fabrication, the V_T can also be tuned by changing the well doping: the available V_T-tuning range was found to be significantly larger that of UTSOI technology. Concerning threshold voltage variability in IFQW pFETs, the V_T mismatch

Table 6.1 Processing details and key analysis results

Wafer nb.	W1	W2	W3	W4
300 mm, blanket Si wafer	×	×	×	×
n-well	×	×	×	×
epi: 4 nm $Si_{0.55}Ge_{0.45}$	×	×		
epi: 4 nm $Si_{0.35}Ge_{0.65}$			×	×
epi: 6 nm Si cap	×	×	×	×
Spike anneal 950 °C		×		×
Ge peak % (SIMS)	45 %	42 %	54 %	46 %

performance was analyzed. The lower well doping of this technology, as compared to bulk Si, was shown to be greatly beneficial in reducing V_T variability.

6.6 SiGe Quantum Well Diffusion Study

As explained in the previous sections, the core feature of the IFQW pFET technology is the SiGe Quantum Well channel, in which the charge carriers are confined. As a result, the exact morphology of this SiGe layer is expected to have a profound influence on the transistors' electrical characteristics. Therefore, this section will investigate the SiGe QW layer in more detail.

6.6.1 Experimental Details

Blanket bulk Si, 300 mm diameter wafers were implanted and annealed with phosphorus and arsenic, creating a deep n-well with a 2×10^{17} cm^{-3} doping concentration. Then, on a first set of samples, a $Si_{0.55}Ge_{0.45}$ channel layer was grown using silane and germane precursor gasses. On a second set of samples, a $Si_{0.35}Ge_{0.65}$ channel layer was grown, using a dichlorosilane and germane chemistry and a lower temperature. On all samples, the SiGe was capped in-situ with a Si cap layer (approximately 6 nm thick). Finally, one wafer from each set received a 950 °C spike anneal. Note that the Ge concentrations mentioned in this paragraph are targeted Ge concentrations—the actual Ge% are different, as stated in the next paragraphs. An overview of the processing is given in Table 6.1.

6.6.2 Physical Analysis

The resulting SiGe layers were analyzed using Secondary Ion Mass Spectroscopy (SIMS). Figure 6.20(a) shows the Ge concentration profile as a function of depth,

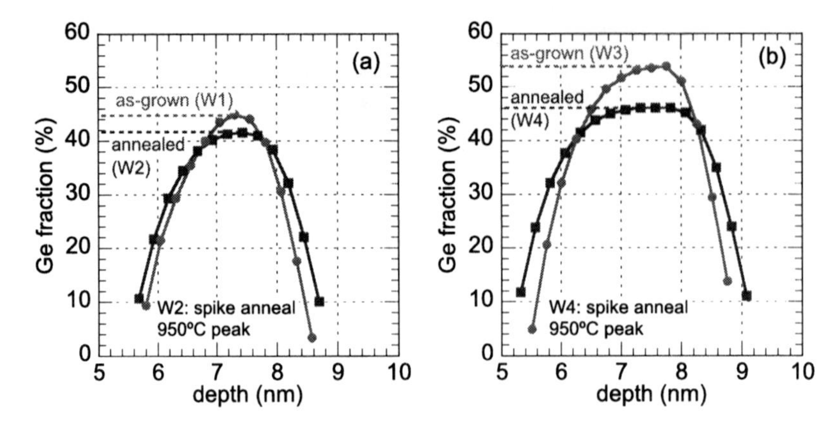

Fig. 6.20 Chemical Ge concentration profiles as a function of depth for two SiGe layers, as-grown and after a spike anneal (peak temperature 950 °C). Notice the rather similar peak Ge concentration after the anneal for both layers. (**a**) Nominal $Si_{0.55}Ge_{0.45}$, (**b**) nominal $Si_{0.35}Ge_{0.65}$

as-grown and after the spike anneal. As can be seen, the peak Ge concentration is on-target at 45 %, while the layer thickness is slightly thinner than designed (3 nm instead of 4 nm). Comparing the as-implanted Ge profile to that after the 950 °C spike anneal, learns that the Ge QW diffusion is rather limited: the peak Ge concentration drops only a few percent to 42 % and the QW layer thickness remains almost unchanged. However, when the target Ge concentration is increased to 65 %, the situation is different. In these samples, the as-grown SiGe layer has a peak Ge concentration of only 54 %. After the spike anneal, the peak Ge concentration drops significantly to 46 %. At the same time, layer thickness is increased by 0.8 nm as a result of this Ge diffusion (Fig. 6.20(b)).

Investigating the Ge% profiles in more detail, two particular observations can be made:

- The peak Ge concentration in the samples after anneal is quite similar, even though the initial Ge% difference of about 10 %.
- The shape of the profiles after the spike anneal resembles that of a rectangular box, with rather well-defined transitions to the Si substrate and the Si cap.

These indicate that the Ge diffusion varies in function of the local Ge fraction. Using the strain-dependent Si-Ge interdiffusion model proposed by Xia et al. [158], the maximum allowed thermal budget for a thin QW channel with a given Ge concentration can be estimated. In that work, an expression is derived for the Ge diffusion constant, depending on the Ge concentration and the local strain. As indicated in Fig. 6.21, this diffusion constant increases quickly as a function of Ge% (layers are assumed to be fully strained (defect-free, relaxation-free) on a Si substrate). Considering the 950 °C spike anneal in our experiments, the local diffusion constant is predicted to increase by almost three orders of magnitude by increasing the Ge fraction from 45 to 65 %. This model can explain our observations: in areas with a high Ge fraction, the Ge atoms are very mobile by the combined effect of Ge

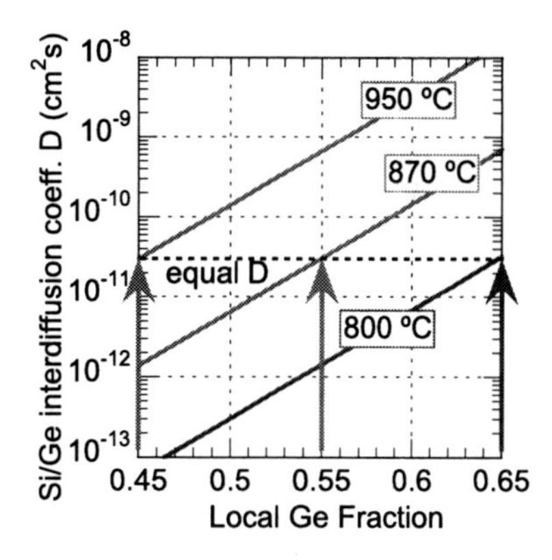

Fig. 6.21 Si/Ge interdiffusion constant D, as calculated using the model from [158] at three different temperatures. Notice the strong increase in diffusivity if the Ge fraction increases. Lowering the temperature to 870 °C and 800 °C results in equal D for Ge fractions of resp. 55 % and 65 % (compared to a 45 % Ge fraction reference)

concentration and compressive strain. Ge diffuses from there to areas with a lower Ge%, lowering the Ge concentration and partially relieving the compressive strain. This, in turn, greatly reduces the diffusion constant. The result of this is indeed the box-like Ge% profile observed in our samples.

An important conclusion is that following such a diffusion model, the peak Ge concentration after the spike anneal is mainly dependent on the temperature of that anneal. Based on our observation that a 950 °C spike anneal results in minimal diffusion for the $Si_{0.55}Ge_{0.45}$ QW channel, maximum spike anneal temperatures can be estimated for other Ge concentrations (requiring an identical Ge diffusion constant as a function of Ge%). As shown in Fig. 6.21, lowering the spike anneal temperature to 870 °C is expected to yield equal diffusion for a $Si_{0.45}Ge_{0.55}$ channel; an even lower spike anneal temperature (800 °C) is required to keep a $Si_{0.35}Ge_{0.65}$ channel intact. This reduced thermal stability of the SiGe channel layers in a SiGe IFQW pFET technology can turn out to be an important limitation in future integration flows where one might want to increase the Ge% in the channel.

6.6.3 Conclusions

In this section, the thermal stability of a SiGe Quantum Well channel was investigated on blanket wafers using SIMS analysis. The observed diffusion was found to be consistent with the Ge-concentration dependent interdiffusion model of Xia et al. [158] is verified. The $Si_{0.55}Ge_{0.45}$ Quantum Well channel (relevant for the *first-* and *second-generation* IFQW pFETs discussed in this chapter), was found to be thermally stable up to 950 °C (spike anneal temperature). Following this anneal, limited Si-Ge interdiffusion still yields a QW channel Ge concentration of 42 %. QW channels with a higher Ge content were observed to require a lower thermal budget to avoid significant Si-Ge interdiffusion. According to the model's predictions, a maximum spike anneal temperature of 800 °C is required to avoid excessive interdiffusion for $Si_{0.35}Ge_{0.65}$ QW channels.

6.7 Conclusions

Following up on the TCAD simulations of the SiGe Implant-Free Quantum Well pFET presented in the previous chapter, the process development to fabricate such devices was discussed. In a first section *first-generation* SiGe IFQW pFETs with raised source/drains were electrically analyzed. Fabricated devices with gate lengths L_G down to 30 nm showed excellent short channel control with *DIBL* and *SS* values of 126 mV/V and 80 mV/dec respectively. Compared to Si control pFETs (identical gatestack), a 50 % higher saturation drive current was obtained.

Leading to *second-generation* SiGe IFQW pFETs, embedded source/drain stressors were integrated into the IFQW pFET architecture. Following calibrated TCAD simulations, a significant boost in intrinsic channel mobility (+~90 %) and reduced series resistance R_{EXT} (−~75 %) was confirmed experimentally. These *second-generation* IFQW pFETs were shown to be competitive with a $L_G = 32$ nm state-of-the-art planar strained-Si technology [153]. Excellent short channel control was maintained in these devices despite a large increase in saturation drive current ($I_{ON} = 1$ mA/μm at $I_{OFF} = 100$ nA/μm and $V_{DD} = 1$ V). Increased performance at lower operating voltage was also demonstrated and comparison with strained-SOI Technology nFETs showed competitive performance. Considering that there is still significant room for further improvement of the IFQW pFET, this comparison suggests that it should be considered a viable device option for upcoming technology nodes.

In a fourth section, the matching performance and V_T-tuning capabilities of IFQW pFETs were investigated. Significant V_T-tuning capabilities were observed, both during circuit operation using body biassing and at the time of fabrication by changing the n-well doping concentration. V_T mismatch performance was also analyzed, showing markedly improved matching, compared to optimized planar bulk Si technologies.

Finally, the thermal stability of a SiGe Quantum Well channel was investigated. While the thermal budget during transistor fabrication should be limited, a $Si_{0.55}Ge_{0.45}$ QW channel was shown to be minimally impacted by a 950 °C spike anneal. Confirming the Ge-concentration dependent interdiffusion model of Xia et al. [158], QW channels with a higher Ge% were shown to require a lower thermal budget, in order to avoid broadening of the SiGe channel.

Chapter 7
Conclusions Future Work and Outlook

High-mobility semiconductors such as germanium and III-V materials offer great possibilities to improve transistor characteristics. However, transistors have to be redesigned in order to fully benefit from these alternative materials. Quantum Well based transistors have been found to be very well suited, as they confine the charge carriers to the high-mobility material using heterostructure isolation. The SiliconGermanium Implant-Free Quantum Well pFET was fabricated using industry-standard infrastructure and showed remarkable short-channel performance. These prototypes were found to be competitive with a $L_G = 32$ nm state-of-the-art planar strained-Si technology, making this a promising transistor architecture for future technology nodes.

7.1 Conclusions

In Chap. 2, the fabrication of shallow junctions in germanium substrates was investigated. Considering p-type junctions, boron and gallium were studied, focussing on electrical activation and diffusion behavior. Under certain conditions, both species deliver a high electrical activation ($N_{ACT} \approx 4 \times 10^{20}$ cm^{-3}), while dopant diffusion remains negligible. As such, it seems likely that the sheet resistance of boron and gallium p-type junctions in germanium can be 3–4 times lower than the ITRS targets for current and upcoming technology nodes, ultimately leading to an increased drive current for aggressively scaled germanium-based technologies. Considering n-type junctions, arsenic was studied, focussing on millisecond laser annealing in an attempt to reduce the concentration-enhanced diffusion and resulting arsenic deactivation commonly observed using classical activation anneals. While sufficiently high active arsenic concentration levels are obtained, significant diffusion is still observed. Consequently, further improvements will be required to meet the ITRS targets for n-type junctions. A direct result is that highly doped p-type junctions, required for germanium pMOSFETs, are available. The lower activation and significant diffusion of their n-type counterparts makes fabrication of scaled germanium nMOSFETs more cumbersome.

G. Hellings, K. De Meyer, *High Mobility and Quantum Well Transistors*,
Springer Series in Advanced Microelectronics 42, DOI 10.1007/978-94-007-6340-1_7,

In Chap. 3, a Monte Carlo simulator was calibrated to enable TCAD simulations of ion implants into germanium substrates. Simulated as-implanted profiles for boron, phosphorus, gallium and arsenic showed good agreement with experimental data, obtained using SIMS measurements. This calibrated MC simulator allows reliable simulations of as-implanted doping profiles and resulting amorphization depth for common dopants in germanium crystals. Using this calibrated MC simulator, the ion implant steps required for a scaled $L_G = 70$ nm germanium pMOSFET technology were designed. Using these, high-performance germanium pMOSFETs were fabricated with physical gate lengths down to 70 nm. Benchmarking of these germanium pFETs shows the potential of germanium to outperform Si as a pFET channel material well into the sub-100 L_G regime. Notably, the 70 nm devices outperform the ITRS requirements for the corresponding technology node concerning saturation drive current I_{ON} by 50 % (maintaining similar I_{OFF} measured at the source). In addition, these germanium pFETs were shown to provide the required performance with a 40 % reduction in active power dissipation, owing to V_{DD} scaling.

In Chap. 4, a TCAD device simulator was extended to allow electrical simulations of germanium pMOSFETs. Specifically, models for carrier mobility and generation/recombination processes were calibrated based on experimental data. Using the calibrated ion implant simulator and the device simulator, electrical simulations of germanium pMOSFETs with L_G ranging from 70 nm to 1 μm were found to be in good agreement with measured I–V curves. Typical transistor performance metrics (I_{ON}, I_{OFF}, *DIBL*, drain-to-substrate leakage, ...) on simulated pFETs were within 5–10 % of the experimental values. Complementing experimental work, this TCAD combination allows optimizing and predicting the performance of new, scaled germanium based devices. Building on these TCAD capabilities, a methodology was presented allowing to study and predict the effect of interface traps in a germanium technology on transistor performance. Finally, the impact of interface traps on MOSFET drive current was investigated. In germanium MOSFETs with SiH_4 passivation, it was found that the electrostatic degradation (due to charging/decharging of traps) accounts for only 20 % of the observed change in drive current I_{ON}. Additional scattering processes (reducing carrier mobility in the channel) were found to be dominant.

In Chap. 5, scaling issues and short channel effects in the bulk MOSFET were analyzed using TCAD simulations. The drain-to-bulk junction leakage in a bulk $L_G = 65$ nm germanium MOSFET technology impose an upper limit on the supply voltage. Considering that smaller gate lengths would require heavier halo implants, it seems unlikely that a planar bulk germanium MOSFET technology would be well suited for future technology nodes ($L_G < 20$ nm). As a result, integrating high-mobility materials such as germanium into future technology nodes will require alternative transistor architectures. Therefore, a class of transistors was introduced, where charge carriers are confined to a Quantum Well (QW) by means of heterostructure confinement. The Implant-Free Quantum Well FET was presented and its scaling performance analyzed for gate lengths down to 16 nm using the Si/SiGe material system as an example, since it combines a certain level of processing maturity (Si-compatibility) without too much lattice mismatch (as compared to a pure germanium QW on a Si substrate). Even at this small gate length, simulated *DIBL* and *SS* remained quite low at 150 mV/V and 90 mV/dec respectively, markedly lower than for equivalent bulk Si pFETs. Zooming in on the critical interface between the QW channel and the underlying substrate, the role of the valence band offset was investigated in the SiGe IFQW pFET. Finally, an InGaAs/GaAs IFQW nFET was also introduced starting from a classical High Electron Mobility Transistor (HEMT), showing good short channel control at a gate length of $L_G = 10$ nm.

In Chap. 6, following up on the TCAD simulations, the process development to fabricate SiGe-based IFQW pFETs was discussed. *First-generation* SiGe IFQW pFETs with raised source/drains were fabricated and electrically analyzed. Devices with gate lengths L_G down to 30 nm showed excellent short channel control with *DIBL* and *SS* values of 126 mV/V and 80 mV/dec respectively. Compared to Si control pFETs (identical gatestack), a 50 % higher saturation drive current was obtained. This confirms the enhanced scalability of IFQW devices. Integrating embedded $Si_{0.75}Ge_{0.25}$ source/drain stressors into this IFQW pFET architecture, *Second-generation* SiGe IFQW pFETs were fabricated and found to yield a significant boost in intrinsic channel mobility and reduced series resistance. These *second-generation* IFQW pFETs were shown to be competitive with a $L_G = 32$ nm state-of-the-art planar strained-Si technology [153]. Excellent short channel control was maintained in these devices despite a large increase in saturation drive current ($I_{ON} = 1$ mA/μm at $I_{OFF} = 100$ nA/μm and $V_{DD} = 1$ V). Increased performance at lower operating voltage was also demonstrated and comparison with strained-SOI technology nFETs showed competitive performance. Considering that there is still significant room for further improvement of the IFQW pFET, this comparison suggests that it should be considered a viable device option for upcoming technology nodes. Finally, the matching performance and V_T-tuning capabilities of IFQW pFETs were investigated, showing significant V_T-tuning capabilities and markedly improved V_T matching, compared to optimized planar bulk Si technologies.

7.2 Future Work and Outlook

At the end of this Ph.D., some recommendations for future research in this field can be made.

1. In this work, the integration of high-mobility materials into VLSI logic circuits was investigated. One of the important conclusions was that to really obtain competitive scaled technologies, dedicated transistor architectures are required. In practice, this probably means such transistors should include strain-engineered channels, advanced high-k dielectrics, metal gate electrodes and other performance boosters found in today's Si-based technologies. One such transistor structure—The Implant-Free Quantum Well FET—was evaluated in this Ph.D. While the first performance evaluation using the Si/Ge material system was promising, future work is needed to further develop the IFQW pFET.

2. Integrating different materials in the same device gives rise to complex mechanical stress and strain profiles. The effect on valence and conduction band energy levels and carrier mobility of the different materials is of great importance for future developments in this field.
3. Since a VLSI technology requires nFETs, as well as pFETs, a second chunk of future work will focus on finding an nFET to be placed next to the IFQW pFET. In the final chapter of this work, the IFQW pFET's performance was found to be comparable to that of strained Silicon-On-Insulator nFETs. Undoubtedly, these two would make an interesting combination, with equally-sized nFETs and pFETs. Alternatively, a full IFQW CMOS technology can be envisioned. A variety of material systems could be used for the nFET, including many III/V based systems. InGaAs-channel based IFQW nFETs in particular [10] are an interesting option. These materials need to be studied and TCAD simulators need to be extended to allow more accurate TCAD simulations of such transistors. And finally, inventive processing solutions need to be developed, co-integrating III/V materials and SiGe/Ge to end up with a true IFQW CMOS technology.
4. With Intel announcing in 2011 [71] that it will start using bulk finFETs for the 22 nm node, it seems that the semiconductor industry is finally implementing more scalable transistor architectures into its products. As this work has focussed on a planar IFQW architecture, a logical extension would be to investigate how the methods described in this work can be applied to a finFET architecture.

Finally, designing transistor structures for upcoming technology nodes has been the focus of this book. Following Gordon Moore's famous extrapolation [103], typical transistor gate lengths will be approaching 10 nm soon. It is still quite unclear how transistors will look like beyond that point. And some predict transistor scaling will finally come to an end around that point. However, there is no real reason to doubt that groundbreaking research, continued innovation and—most of all—bright scientists and engineers will succeed in finding answers. Scaling will continue, for now....

References

1. T. Ando, M.M. Frank, K. Choi, C. Choi, J. Bruley, M. Hopstaken, M. Copel, E. Cartier, A. Kerber, A. Callegari, D. Lacey, S. Brown, Q. Yang, V. Narayanan, Understanding mobility mechanisms in extremely scaled HfO_2 (EOT 0.42 nm) using remote interfacial layer scavenging technique and Vt-tuning dipoles with gate-first process, in *IEEE International Electron Devices Meeting* (2009), pp. 1–4
2. F. Andrieu, O. Weber, J. Mazurier, O. Thomas, J.-P. Noel, C. Fenouillet-Beandranger, J.-P. Mazellier, P. Perreau, T. Poiroux, Y. Morand, T. Morel, S. Allegret, V. Loup, S. Barnola, F. Martin, J.-F. Damlencourt, I. Servin, M. Casse and, X. Garros, O. Rozeau, M.-A. Jaud, G. Cibrario, J. Cluzel, A. Toffoli, F. Allain, R. Kies, D. Lafond, V. Delaye, C. Tabone, L. Tosti, L. Breandvard, P. Gaud, V. Paruchuri, K.K. Bourdelle, W. Schwarzenbach, O. Bonnin, B.-Y. Nguyen, B. Doris, F. Bœanduf, T. Skotnicki, O. Faynot, Low leakage and low variability Ultra-Thin Body and Buried Oxide (UT2B) SOI technology for 20nm low power CMOS and beyond, in *Symposium on VLSI Technology* (2010), pp. 57–58
3. D.A. Antoniadis, A. Khakifirooz, MOSFET performance scaling: limitations and future options, in *International Electron Devices Meeting* (2008), pp. 1–4
4. A. Asenov, K. Samsudin, Variability in nanoscale UTB SOI devices and its impact on circuits and systems, in *Nanoscaled Semiconductor-On-Insulator Structures and Devices*, vol. 17 (2007), pp. 259–302
5. D.G. Ashworth, R. Oven, B. Mundin, Representation of ion implantation profiles by Pearson frequency distribution curves. J. Phys. D, Appl. Phys. **23**(7), 870 (1990)
6. E. Baravelli, M. Jurczak, N. Speciale, K. De Meyer, A. Dixit, Impact of LER and random dopant fluctuations on FinFET matching performance. IEEE Trans. Nanotechnol. **7**(3), 291–298 (2008)
7. F. Bellenger, M. Houssa, A. Delabie, V. Afanasiev, T. Conard, M. Caymax, M. Meuris, K. De Meyer, M.M. Heyns, Passivation of $Ge(100)/GeO_2/high\text{-}\kappa$ gate stacks using thermal oxide treatments. J. Electrochem. Soc. **155**(2), G33–G38 (2008)
8. B. Benbakhti, J.S. Ayubi-Moak, K. Kalna, D. Lin, G. Hellings, G. Brammertz, K. De Meyer, I. Thayne, A. Asenov, Impact of interface state trap density on the performance characteristics of different III-V MOSFET architectures. Microelectron. Reliab. **50**(3), 360–364 (2010)
9. B. Benbakhti, K. Kalna, K. Chan, A. Asenov, G. Hellings, G. Eneman, K. De Meyer, M. Meuris, Design and analysis of a new $In_{0.53}Ga_{0.47}As$ implant-free quantum-well device structure, in *European MRS Meeting. Symposium H* (2010)
10. B. Benbakhti, E. Towie, K. Kalna, G. Hellings, G. Eneman, K. De Meyer, M. Meuris, A. Asenov, Monte Carlo analysis of $In_{0.53}Ga_{0.47}As$ implant-free quantum-well device performance, in *Silicon Nanoelectronics Workshop Proc.* (2010), pp. 17–18
11. B. Benbakhti, K. Chan, E. Towie, K. Kalna, C. Riddet, X. Wang, G. Eneman, G. Hellings, K. De Meyer, M. Meuris, A. Asenov, Numerical analysis of the new implant-free quantum-well

G. Hellings, K. De Meyer, *High Mobility and Quantum Well Transistors*,
Springer Series in Advanced Microelectronics 42, DOI 10.1007/978-94-007-6340-1,

CMOS—DualLogic approach. Solid-State Electron. **63**(1), 14–18 (2011)
12. B. Benbakhti, K. Kalna, K. Chan, E. Towie, G. Hellings, G. Eneman, K. De Meyer, M. Meuris, A. Asenov, Design and analysis of the $In_{0.53}Ga_{0.47}As$ implant-free quantum-well device structure. Microelectron. Eng. **88**(4), 358–361 (2011)
13. B. Benbakhti, A. Martinez, K. Kalna, G. Hellings, G. Eneman, K. De Meyer, M. Meuris, Simulation study of performance for a 20nm gate length $In_{0.53}Ga_{0.47}As$ implant free quantum well MOSFET. IEEE Trans. Nanotechnol. **11**, 808–817 (2012)
14. S. Brotzmann, H. Bracht, Intrinsic and extrinsic diffusion of phosphorus, arsenic, and antimony in germanium. J. Appl. Phys. **103**(3), 033508 (2008)
15. D.P. Brunco, B. De Jaeger, G. Eneman, A. Satta, V. Terzieva, L. Souriau, F.E. Leys, G. Pourtois, M. Houssa, K. Opsomer, G. Nicholas, M. Meuris, M.M. Heyns, Germanium: the past and possibly a future material for microelectronics. ECS Trans. **11**(4), 479–493 (2007)
16. D.P. Brunco, B. De Jaeger, G. Eneman, J. Mitard, G. Hellings, A. Satta, V. Terzieva, L. Souriau, F.E. Leys, G. Pourtois, M. Houssa, G. Winderickx, E. Vrancken, S. Sioncke, K. Opsomer, G. Nicholas, M. Caymax, A. Stesmans, J. Van Steenbergen, P.W. Mertens, M. Meuris, M.M. Heyns, Germanium MOSFET devices: advances in materials understanding, process development, and electrical performance. J. Electrochem. Soc. **155**(7), H552–H561 (2008)
17. D.M. Caughey, R.E. Thomas, Carrier mobilities in silicon empirically related to doping and field. Proc. IEEE **55**(12), 2192–2193 (1967)
18. M. Caymax, G. Eneman, F. Bellenger, C. Merckling, A. Delabie, G. Wang, R. Loo, E. Simoen, J. Mitard, B. De Jaeger, G. Hellings, K. De Meyer, M. Meuris, M. Heyns, Germanium for advanced CMOS anno 2009: a SWOT analysis, in *IEEE International Electron Devices Meeting (IEDM)* (2009), pp. 461–464
19. R. Chau, S. Datta, M. Doczy, B. Doyle, B. Jin, J. Kavalieros, A. Majumdar, M. Metz, M. Radosavljevic, Benchmarking nanotechnology for high-performance and low-power logic transistor applications. IEEE Electron Device Lett. **4**(2), 153–158 (2005)
20. T.-C. Chen, Challenges for silicon technology scaling in the nanoscale era, in *European Solid State Circuits Research Conference* (2009), pp. 1–7
21. K. Cheng, A. Khakifirooz, P. Kulkarni, S. Kanakasabapathy, S. Schmitz, A. Reznicek, T. Adam, Y. Zhu, J. Li, J. Faltermeier, T. Furukawa, L.F. Edge, B. Haran, S.-C. Seo, P. Jamison, J. Holt, X. Li, R. Loesing, Z. Zhu, R. Johnson, A. Upham, T. Levin, M. Smalley, J. Herman, M. Di, J. Wang, D. Sadana, P. Kozlowski, H. Bu, B. Doris, J. O'Neill, Fully depleted extremely thin SOI technology fabricated by a novel integration scheme featuring implant-free, zero-silicon-loss, and faceted raised source/drain, in *Symposium on VLSI Technology* (2009), pp. 212–213
22. A. Chroneos, R.W. Grimes, B.P. Uberuaga, S. Brotzmann, H. Bracht, Vacancy-arsenic clusters in germanium. Appl. Phys. Lett. **91**(19), 192106 (2007)
23. C.-O. Chui, H. Kim, D. Chi, B.B. Triplett, P.C. McIntyre, K.C. Saraswat, A sub-400 deg°C germanium MOSFET technology with high-κ dielectric and metal gate, in *International Electron Devices Meeting* (2002), pp. 437–440
24. C.-O. Chui, L. Kulig, J. Moran, W. Tsai, K.C. Saraswat, Germanium n-type shallow junction activation dependences. Appl. Phys. Lett. **87**(9), 091909 (2005)
25. T. Clarysse, D. Vanhaeren, I. Hoflijk, W. Vandervorst, Characterization of electrically active dopant profiles with the spreading resistance probe. Mater. Sci. Eng., R Rep. **47**(5–6), 123–206 (2004)
26. T. Clarysse, P. Eyben, T. Janssens, I. Hoflijk, D. Vanhaeren, A. Satta, M. Meuris, W. Vandervorst, J. Bogdanowicz, G. Raskin, Active dopant characterization methodology for germanium. J. Vac. Sci. Technol., B **24**(1), 381–389 (2006)
27. J.P. Colinge, *Silicon-On-Insulator Technology: Materials to VLSI* (Kluwer Academic, Boston, 1991)
28. J.-P. Colinge, *FinFETs and Other Multi-Gate Transistors*, 1st edn. (Springer, Berlin, 2007)
29. L. Csepregi, R. Kullen, J. Mayer, T. Sigmon, Regrowth kinetics of amorphous Ge layers created by ^{74}Ge and ^{27}Si implantation of Ge crystals. Solid State Commun. **21**(11), 1109

(1977)
30. B. De Jaeger, R. Bonzom, F. Leys, J. Steenbergen, G. Winderickx, E. Van Moorhem, G. Raskin, F. Letertre, T. Billon, M. Meuris, M. Heyns, Optimisation of a thin epitaxial Si layer as a Ge passivation layer to demonstrate deep sub-micron n- and p-FETs on Ge-On-insulator substrates. Microelectron. Eng. **80**, 26–29 (2005)
31. B. De Jaeger, G. Nicholas, D.P. Brunco, G. Eneman, M. Meuris, M.M. Heyns, High performance high-k/metal gate Ge pMOSFETs with gate lengths down to 125 nm and halo implant, in *37th European Solid State Device Research Conference* (2007), pp. 462–465
32. G. Declerck, A look into the future of nanoelectronics, in *Symposium on VLSI Technology* (2005), pp. 6–10
33. A. Dixit, FEOL CMOS process-and device-parasitics IN SOI MuGFETs. Ph.D. Dissertation, KU Leuven (2007)
34. J.M. Dorkel, Ph. Leturcq, Carrier mobilities in silicon semi-empirically related to temperature, doping and injection level. Solid-State Electron. **24**(9), 821–825 (1981)
35. G. Du, X.Y. Liu, Z.-L. Xia, Y.K. Wang, D.Q. Hou, J.F. Kang, R.Q. Han, Evaluations of scaling properties for Ge on insulator MOSFETs in nano-scale. Jpn. J. Appl. Phys. **44**(4B), 2195–2197 (2005)
36. R. Duffy, M. Shayesteh, M. White, J. Kearney, A.-M. Kelleher, The formation, stability, and suitability of n-type junctions in germanium formed by solid phase epitaxial recrystallization. Appl. Phys. Lett. **96**(23), 231909 (2010)
37. G. Eneman, Design, fabrication and characterizatino of advanced field effect transistors with strained silicon channels. Ph.D. Dissertation, KU Leuven (2006)
38. G. Eneman, M. Wiot, A. Brugere, O.S.I. Casain, S. Sonde, D.P. Brunco, B. De Jaeger, A. Satta, G. Hellings, K. De Meyer, C. Claeys, M. Meuris, M.M. Heyns, E. Simoen, Impact of donor concentration, electric field, and temperature effects on the leakage current in germanium p+/n junctions. IEEE Trans. Electron Devices **55**(9), 2287–2296 (2008)
39. G. Eneman, B. De Jaeger, E. Simoen, D.P. Brunco, G. Hellings, J. Mitard, K. De Meyer, M. Meuris, M.M. Heyns, Quantification of drain extension leakage in a scaled bulk germanium pMOS technology. IEEE Trans. Electron Devices **56**(12), 3115–3122 (2009)
40. G. Eneman, G. Hellings, J. Mitard, L. Witters, S. Yamaguchi, M. Garcia Bardon, P. Christie, C. Ortolland, A. Hikavyy, P. Favia, M. Bargallo Gonzalez, E. Simoen, F. Crupi, M. Kobayashi, J. Franco, S. Takeoka, R. Krom, H. Bender, R. Loo, C. Claeys, K. De Meyer, T. Hoffmann, $Si_{1-x}Ge_x$-channel PFETs: scalability, layout considerations and compatibility with other stress techniques, in *Dielectrics in Nanosystems and Graphene, Ge/III-V, Nanowires and Emerging Materials for Post-CMOS Applications*, vol. 3 (2011), pp. 493–503
41. P. Eyben, M. Xu, N. Duhayon, T. Clarysse, S. Callewaert, W. Vandervorst, Scanning spreading resistance microscopy and spectroscopy for routine and quantitative two-dimensional carrier profiling. J. Vac. Sci. Technol., B **20**(1), 471–478 (2002)
42. P. Favia, M. Bargallo Gonzales, E. Simoen, P. Verheyen, D. Klenov, H. Bender, Nanobeam diffraction: technique evaluation and strain measurement on complementary metal oxide semiconductor devices. J. Electrochem. Soc. **158**(4), H438–H446 (2011)
43. O. Faynot, F. Andrieu, O. Weber, C. Fenouillet-Beandranger, P. Perreau, J. Mazurier, T. Benoist, O. Rozeau, T. Poiroux, M. Vinet, L. Grenouillet, J.-P. Noel, N. Posseme, S. Barnola, F. Martin, C. Lapeyre, M. Casse and, X. Garros, M.-A. Jaud, O. Thomas, G. Cibrario, L. Tosti, L. Brevard, C. Tabone, P. Gaud, S. Barraud, T. Ernst, S. Deleonibus, Planar fully depleted soi technology: a powerful architecture for the 20nm node and beyond, in *International Electron Devices Meeting* (2010)
44. V.I. Fistul, M.I. Iglitsyn, E.M. Omelyanovskii, Mobility of electrons in germanium strongly doped with arsenic. Sov. Phys., Solid State **4**(4), 784–785 (1962)
45. D. Fleury, A. Cros, K. Romanjek, D. Roy, F. Perrier, B. Dumont, H. Brut, G. Ghibaudo, Automatic extraction methodology for accurate measurements of effective channel length on 65-nm MOSFET technology and below. IEEE Trans. Semicond. Manuf. **21**(4), 504–512 (2008)

46. J.G. Fossum, D.S. Lee, A physical model for the dependence of carrier lifetime on doping density in nondegenerate silicon. Solid-State Electron. **25**, 741–747 (1982)
47. E. Gaubas, M. Bauza, A. Uleckas, J. Vanhellemont, Carrier lifetime studies in Ge using microwave and infrared light techniques. Mater. Sci. Semicond. Process. **9**(4–5), 781–787 (2006). Also in Proceedings of Symposium T E-MRS 2006 Spring Meeting on Germanium Based Semiconductors from Materials to Devices
48. T. Ghani, M. Armstrong, C. Auth, M. Bost, P. Charvat, G. Glass, T. Hoffmann, K. Johnson, C. Kenyon, J. Klaus, B. McIntyre, K. Mistry, A. Murthy, J. Sandford, M. Silberstein, S. Sivakumar, P. Smith, K. Zawadzki, S. Thompson, M. Bohr, A 90nm high volume manufacturing logic technology featuring novel 45nm gate length strained silicon CMOS transistors, in *IEEE International Electron Devices Meeting* (2003), pp. 11.6.1–11.6.3
49. J. Glassbrenner, G.A. Slack, Thermal conductivity of silicon and germanium from 3K to the melting point. Phys. Rev. **164**(4A), 1058–1069 (1964)
50. O.A. Golikova, B.Ya. Moizhes, L.S. Stil'bans, Hole mobility of germanium as a function of concentration and temperature. Sov. Phys., Solid State **3**(10), 2259–2265 (1962)
51. V. Heera, A. Mucklich, M. Posselt, M. Voelskow, C. Wundisch, B. Schmidt, R. Skrotzki, K.H. Heinig, T. Herrmannsdorfer, W. Skorupa, Heavily Ga-doped germanium layers produced by ion implantation and flash lamp annealing: structure and electrical activation. J. Appl. Phys. **107**(5), 053508 (2010)
52. G. Hellings, G. Eneman, B. De Jaeger, J. Mitard, K. De Meyer, M. Meuris, M. Heyns, Scalability of quantum well device for digital logic applications, in *Silicon Nanoelectronincs Workshop Proc.* (2009), pp. 33–34
53. G. Hellings, C. Wuendisch, G. Eneman, E. Simoen, T. Clarysse, M. Meuris, W. Vandervorst, M. Posselt, K. De Meyer, Implantation, diffusion, activation, and recrystallization of gallium implanted in preamorphized and crystalline germanium. Electrochem. Solid-State Lett. **12**(12), H417–H419 (2009)
54. G. Hellings, G. Eneman, R. Krom, B. De Jaeger, J. Mitard, A. De Keersgieter, T. Hoffmann, M. Meuris, K. De Meyer, Electrical TCAD simulations of a germanium pMOSFET technology. IEEE Trans. Electron Devices **57**(10), 2539–2546 (2010)
55. G. Hellings, L. Witters, R. Krom, J. Mitard, A. Hikavyy, R. Loo, A. Schulze, G. Eneman, C. Kerner, J. Franco, T. Chiarella, S. Takeoka, J. Tseng, W. Wang, W. Vandervorst, P. Absil, S. Biesemans, M. Heyns, K. De Meyer, M. Meuris, T. Hoffmann, Implant-free SiGe quantum well pFET: a novel, highly scalable and low thermal budget device, featuring raised source/drain and high-mobility channel, in *IEEE International Electron Devices Meeting (IEDM)* (2010), pp. 241–244
56. G. Hellings, G. Eneman, J. Mitard, K. Martens, W.-E. Wang, T. Hoffmann, M. Meuris, K. De Meyer, A fast and accurate method to study the impact of interface traps on germanium MOS performance. IEEE Trans. Electron Devices **58**(4), 938–944 (2011)
57. G. Hellings, G. Eneman, J. Mitard, L. Witters, S. Yamaguchi, A. Hikavyy, P. Favia, K. De Meyer, High-performance SiGe implant-freequantum well pFET technology with raised and embedded source/drain stressors. IEEE Electron Device Lett. (2011). Submitted
58. G. Hellings, A. Hikavyy, J. Mitard, L. Witters, B. Benbakhti, A. Alian, N. Waldron, H. Bender, G. Eneman, R. Krom, R. Loo, M. Heyns, M. Meuris, T. Hoffmann, K. De Meyer, The implant-free quantum well field-effect-transistor: harnessing the power of heterostructures, in *7th International Conference on Si Epitaxy and Heterostructures (ICSI-7)* (2011)
59. G. Hellings, A. Hikavyy, J. Mitard, L. Witters, B. Benbakhti, A. Alian, N. Waldron, H. Bender, G. Eneman, R. Krom, R. Loo, M. Heyns, M. Meuris, T. Hoffmann, K. De Meyer, The implant-free quantum well field-effect-transistor: harnessing the power of heterostructures. Thin Solid Films **520**(8), 3326–3331 (2011)
60. G. Hellings, J. Mitard, R. Krom, L. Witters, G. Eneman, A. Hikavyy, R. Loo, H. Bender, T. Hoffmann, K. De Meyer, Scalability and threshold voltage dependency for the implant-free SiGe quantum well pFET with raised source/drain, in *Silicon Nanoelectronics Workshop* (2011), pp. 5–6

61. G. Hellings, E. Rosseel, T. Clarysse, D.H. Petersen, O. Hansen, P.F. Nielsen, E. Simoen, G. Eneman, B. De Jaeger, T. Hoffmann, K. De Meyer, W. Vandervorst, Systematic study of shallow junction formation on germanium substrates. Microelectron. Eng. **88**(4), 347–350 (2011). Post-Si-CMOS electronic devices: the role of Ge and III-V materials
62. G. Hellings, G. Eneman, M. Meuris, Scalable quantum well device and method for manufacturing the same. US Patent no. 7915608
63. G. Hellings, G. Eneman, M. Meuris, Scalable quantum well device and method for manufacturing the same. European Patent Office Application no. EP2120266
64. M. Heyns, A. Alian, G. Brammertz, M. Caymax, Y. Chang, L. Chu, B. De Jaeger, G. Eneman, F. Gencarelli, G. Groeseneken, G. Hellings, A. Hikavyy, T. Hoffmann, M. Houssa, C. Huyghebaert, D. Leonelli, D. Lin, R. Loo, W. Magnus, C. Merckling, M. Meuris, J. Mitard, L. Nyns, T. Orzali, R. Rooyackers, S. Sioncke, B. Soree, X. Sun, A. Vandooren, A. Verhulst, B. Vincent, N. Waldron, G. Wang, W. Wang, L. Witters, Advancing CMOS beyond the Si roadmap with Ge and III/V devices, in *IEEE International Electron Devices Meeting (IEDM)* (2011), pp. 299–302
65. D.P. Hickey, Z.L. Bryan, K.S. Stones, R.G. Elliman, E.E. Haller, Defects in Ge and Si caused by 1 MeV Si^+ implantation. J. Vac. Sci. Technol., B **26**(1), 425–429 (2008)
66. A. Hikavyy, R. Loo, L. Witters, S. Takeoka, J. Geypen, B. Brijs, C. Merckling, M. Caymax, J. Dekoster, SiGe SEG growth for buried channels p-MOS devices. ECS Trans. **25**(7), 201–210 (2009)
67. W.K. Hofker, D.P. Oosthoek, N.J. Koeman, H.A.M. de grefte, Concentration profiles of boron implantations in amorphous and polycrystalline silicon. Philips Res. Rep. **24**(4), 223–231 (1975)
68. J. Huang, N. Wu, Q. Zhang, C. Zhu, A.A.O. Tay, G. Chen, M. Hong, Germanium n^+/p junction formation by laser thermal process. Appl. Phys. Lett. **87**(17), 173507 (2005)
69. G.A.M. Hurkx, D.B.M. Klaassen, M.P.G. Knuvers, A new recombination model for device simulation including tunneling. IEEE Trans. Electron Devices **39**(2), 331–338 (1992)
70. G. Impellizzeri, S. Mirabella, E. Bruno, A.M. Piro, M.G. Grimaldi, B activation and clustering in ion-implanted Ge. J. Appl. Phys. **105**(6), 063533 (2009)
71. Intel corporation, Intel 22nm 3-D Tri-Gate Transistor Technology (2011). http://newsroom.intel.com/docs/DOC-2032.pdf
72. International Technology Roadmap for Semiconductors (ITRS). 2001 edition. Online: http://www.itrs.net
73. International Technology Roadmap for Semiconductors (ITRS). 2007 edition. Online: http://www.itrs.net
74. International Technology Roadmap for Semiconductors (ITRS). 2009 edition. Online: http://www.itrs.net
75. N. Ioannou, D. Skarlatos, N.Z. Vouroutzis, S.N. Georga, C.A. Krontiras, C. Tsamis, Gallium implantation and diffusion in crystalline germanium. Electrochem. Solid-State Lett. **13**(3), H70–H72 (2010)
76. T. Janssens, C. Huyghebaert, D. Vanhaeren, G. Winderickx, A. Satta, M. Meuris, W. Vandervorst, Heavy ion implantation in Ge: dramatic radiation induced morphology in Ge. J. Vac. Sci. Technol., B **24**(1), 510–514 (2006)
77. B.C. Johnson, P. Gortmaker, J.C. McCallum, Intrinsic and dopant-enhanced solid-phase epitaxy in amorphous germanium. Phys. Rev. B **77**(21), 214109 (2008)
78. P. Jourand, Biomedical instrumentation for implantable and warable sensor sytems. Ph.D. Dissertation, KU Leuven (2012)
79. D. Kim, T. Krishnamohan, L. Smith, H.-S.P. Wong, K.C. Saraswat, Band to band tunneling study in high mobility materials: III-V, Si, Ge and strained SiGe, in *65th Annual Device Research Conference* (2007), pp. 57–58
80. T. Krishnamohan, K. Saraswat, High mobility Ge and III-V materials and novel device structures for high performance nanoscale MOSFETS, in *European Solid-State Device Research Conference* (2008), pp. 38–46

81. T. Krishnamohan, Z. Krivokapic, K. Uchida, Y. Nishi, K.C. Saraswat, High-mobility ultrathin strained Ge MOSFETs on bulk and SOI with low band-to-band tunneling leakage: experiments. IEEE Trans. Electron Devices **53**(5), 990–999 (2006)
82. R. Krom, G. Hellings, J. Mitard, L. Witters, A. Hikavyy, G. Eneman, N. Waldron, M. Heyns, T. Hoffmann, K. De Meyer, On the importance of source/drain series resistance in implant-free SiGe quantum well FETs, in *Silicon Nanoelectronics Workshop* (2011), pp. 7–8
83. K. Kuhn, C. Kenyon, A. Kornfeld, M. Liu, A. Maheshwari, W. Shih, S. Sivakumar, G. Taylor, P. Van Der Voorn, K. Zawadzki, Managing process variation in Intel's 45nm CMOS technology. Intel Technol. J. **12**(2), 93–110 (2008)
84. T.Y. Kuo, J.E. Cunningham, E.F. Schubert, W.T. Tsang, T.H. Chiu, F. Ren, C.G. Fonstad, Selectively δ-doped quantum well transistor grown by gas-source molecular-beam epitaxy. J. Appl. Phys. **64**(6), 3324–3327 (1988)
85. R.D. Larrabee, Drift velocity saturation in p-type germanium. J. Appl. Phys. **30**(6), 857–859 (1959)
86. C. Le Royer, L. Clavelier, C. Tabone, C. Deguet, L. Sanchez, J.-M. Hartmann, M.-C. Roure, H. Grampeix, S. Deleonibus, 0.12µm p-MOSFETs with high-k and metal gate fabricated in a Si process line on 200mm GeOI wafers, in *37th European Solid State Device Research Conference* (2007), pp. 458–461
87. C. Le Royer, B. Vincent, L. Clavelier, J.-F. Damlencourt, C. Tabone, P. Batude, D. Blachier, R. Truche, Y. Campidelli, Q.T. Nguyen, S. Cristoloveanu, S. Soliveres, G. Le Carval, F. Boulanger, T. Billon, D. Bensahel, S. Deleonibus, High-k and metal-gate pMOSFETs on GeOI obtained by Ge enrichment: analysis of ON and OFF performances. IEEE Electron Device Lett. **29**(6), 635–637 (2008)
88. S.S. Li, W.R. Thurder, The dopant density and temperature dependence of electron mobility and resistivity in n-type silicon. Solid-State Electron. **20**(7), 609–616 (1977)
89. J.E. Lilienfeld, Method and apparatus for controlling electric current. Canadian Patent Office no. CA272437
90. D. Linten, Evaluation of deep-dub-quarter micron CMOS technology: low noise amplifiers, oscillators and ESD reliability. Ph.D. Dissertation, Vrije Universiteit Brussel (2006)
91. Y. Liu, N. Neophytou, T. Low, G. Klimeck, M.S. Lundstrom, A tight-binding study of the ballistic injection velocity for ultrathin-body SOI MOSFETs. IEEE Trans. Electron Devices **55**(3), 866–871 (2008)
92. C. Lombardi, S. Manzini, A. Saporito, M. Vanzi, A physically based mobility model for numerical simulation of nonplanar devices. IEEE Trans. Comput.-Aided Des. Integr. Circuits Syst. **7**(11), 1164–1171 (1988)
93. R. Loo, M. Caymax, I. Peytier, S. Decoutere, N. Collaert, P. Verheyen, W. Vandervorst, K. De Meyer, Successful selective epitaxial $Si_{1-x}Ge_x$ deposition process for HBT-BiCMOS and high mobility heterojunction pMOS applications. J. Electrochem. Soc. **150**(10), G638–G647 (2003)
94. M.S. Lundstrom, On the mobility versus drain current relation for a nanoscale MOSFET. IEEE Electron Device Lett. **22**(6), 293–295 (2001)
95. K. Martens, J. Mitard, B. De Jaeger, M. Meuris, H. Maes, G. Groeseneken, F. Minucci, F. Crupi, Impact of Si-thickness on interface and device properties for Si-passivated Ge pMOSFETs, in *Solid-State Device Research Conference* (2008), pp. 138–141
96. K. Martens, C. On Chui, G. Brammertz, B. De Jaeger, D. Kuzum, M. Meuris, M. Heyns, T. Krishnamohan, K. Saraswat, H.E. Maes, G. Groeseneken, On the correct extraction of interface trap density of mos devices with high-mobility semiconductor substrates. IEEE Trans. Electron Devices **55**(2), 547–556 (2008)
97. G. Masetti, M. Severi, S. Solmi, Modeling of carrier mobility against carrier concentration in arsenic-, phosphorus-, and boron-doped silicon. IEEE Trans. Electron Devices **30**(7), 764–769 (1983)
98. S. Mirabella, G. Impellizzeri, A.M. Piro, E. Bruno, M.G. Grimaldi, Activation and carrier mobility in high fluence B implanted germanium. Appl. Phys. Lett. **92**, 251909 (2008)

99. J. Mitard, K. Martens, B. De Jaeger, J. Franco, C. Shea, C. Plourde, F. Leys, R. Loo, G. Hellings, G. Eneman, W. Wang, V. Lin, B. Kaczer, K. De Meyer, T. Hoffmann, S. De Gendt, M. Caymax, M. Meuris, M. Heyns, Impact of Epi-Si growth temperature on Ge-pFET performance, in *39th European Solid-State Device Research Conference (ESSDERC)* (2009), pp. 411–414
100. J. Mitard, C. Shea, B. De Jaeger, A. Pristera, G. Wang, M. Houssa, G. Eneman, G. Hellings, W.E. Wang, J.C. Lin, F.E. Leys, R. Loo, G. Winderickx, E. Vrancken, A. Stesmans, K. De Meyer, M. Caymax, L. Pantisano, M. Meuris, M. Heyns, Impact of EOT scaling down to 0.85nm on 70nm GE-pFETs technology with STI, in *Symposium on VLSI Technology* (2009), pp. 82–83
101. J. Mitard, L. Witters, G. Hellings, R. Krom, J. Franco, G. Eneman, A. Hikavyy, B. Vincent, R. Loo, P. Favia, H. Dekkers, E. Altamirano Sanchez, A. Vanderheyden, D. Vanhaeren, P. Eyben, S. Takeoka, S. Yamaguchi, M. Dal Van, W. Wang, S. Hong, W. Vandervorst, K. De Meyer, S. Biesemans, P. Absil, N. Horiguchi, T. Hoffmann, 1mA/μm-ION strained SiGe45-IFQW pFETs with raised and embedded S/D, in *Symposium on VLSI Technology* (2011), pp. 134–135
102. N. Moll, M.R. Hueschen, A. Fischer-Colbrie, Pulse-doped AlGaAs/InGaAs pseudomorphic MODFETs. IEEE Trans. Electron Devices **35**(7), 879–886 (1988)
103. G.E. Moore, Cramming more components onto integrated circuits. Electronics **38**, 8 (1965). http://download.intel.com/museum/Moores_Law/Articles-Press_releases/Gordon_Moore_1965_Article.pdf
104. M.H. Na, E.J. Nowak, W. Haensch, J. Cai, The effective drive current in CMOS inverters, in *International Electron Devices Meeting* (2002), pp. 121–124
105. G. Nicholas, B. De Jaeger, D.P. Brunco, P. Zimmerman, G. Eneman, K. Martens, M. Meuris, M.M. Heyns, High-performance deep submicron Ge pMOSFETs with halo implants. IEEE Electron Device Lett. **54**(9), 2503–2511 (2007)
106. T. Noda, W. Vandervorst, S. Felch, V. Parihar, A. Cuperus, R. Mcintosh, C. Vrancken, E. Rosseel, H. Bender, B. Van Daele, M. Niwa, H. Umimoto, R. Schreutelkamp, P.P. Absil, M. Jurczak, K. De Meyer, S. Biesemans, T.Y. Hoffmann, Analysis of As, P diffusion and defect evolution during sub-millisecond non-melt laser annealing based on an atomistic kinetic Monte Carlo approach, in *International Electron Devices Meeting* (2007), pp. 955–958
107. A.S. Okhotin, A.S. Pushkarskii, V.V. Gorbachev, *Thermophysical Properties of Semiconductors* (Atom, Moscow, 1972)
108. P. Palestri, D. Esseni, S. Eminente, C. Fiegna, E. Sangiorgi, L. Selmi, Understanding quasi-ballistic transport in nano-MOSFETs: part I-scattering in the channel and in the drain. IEEE Trans. Electron Devices **52**(12), 2727–2735 (2005)
109. L. Pantisano, L. Trojman, J. Mitard, B. De Jaeger, S. Severi, G. Eneman, G. Crupi, T. Hoffmann, I. Ferain, M. Meuris, M. Heyns, Fundamentals and extraction of velocity saturation in sub-100nm (110)-Si and (100)-Ge, in *Symposium on VLSI Technology* (2008), pp. 52–53
110. T.P. Pearsall, *GaInAsP Alloy Semiconductors* (Wiley, New York, 1982)
111. M.J.M. Pelgrom, H.P. Tuinhout, M. Vertregt, Transistor matching in analog CMOS applications, in *IEEE International Electron Devices Meeting* (1998), pp. 915–918
112. C.L. Petersen, T.M. Hansen, P. Bøggild, A. Boisen, O. Hansen, T. Hassenkam, F. Grey, Scanning microscopic four-point conductivity probes. Sens. Actuators A, Phys. **96**(1), 53–58 (2002)
113. D.H. Petersen, O. Hansen, R. Lin, P.F. Nielsen, T. Clarysse, J. Goossens, E. Rosseel, W. Vandervorst, High precision micro-scale Hall effect characterization method using in-line micro four-point probes, in *16th IEEE International Conference on Advanced Thermal Processing of Semiconductors (RTP 2008)* (2008), pp. 251–256
114. D.H. Petersen, R. Lin, T.M. Hansen, E. Rosseel, W. Vandervorst, C. Markvardsen, D. Kjæandr, P.F. Nielsen, Comparative study of size dependent four-point probe sheet resistance measurement on laser annealed ultra-shallow junctions. J. Vac. Sci. Technol., B Microelectron. Nanometer Struct. Process. Meas. Phenom. **26**(1), 362–367 (2008)

115. R. Pillarisetty, B. Chu-Kung, S. Corcoran, G. Dewey, J. Kavalieros, H. Kennel, R. Kotlyar, V. Le, D. Lionberger, M. Metz, N. Mukherjee, J. Nah, W. Rachmady, M. Radosavljevic, U. Shah, S. Taft, H. Then, N. Zelick, R. Chau, High mobility strained germanium quantum well field effect transistor as the p-channel device option for low power (Vcc = 0.5 V) III–V CMOS architecture, in *International Electron Devices Meeting* (2010)
116. M. Posselt, Crystal-TRIM and its application to investigations on channeling effects during ion implantation, in *Radiation Effects and Defects in Solids: Incorporating Plasma Science and Plasma Technology*, vol. 130 (1994), pp. 87–119
117. M. Posselt, B. Schmidt, W. Anwand, R. Grötzschel, V. Heera, A. Mücklich, C. Wündisch, W. Skorupa, H. Hortenbach, S. Gennaro, M. Bersani, D. Giubertoni, A. Möller, H. Bracht, P implantation into preamorphized germanium and subsequent annealing: solid phase epitaxial regrowth, P diffusion, and activation, in *International Workshop on Insight in Semiconductor Device Fabrication, Metrology and Modeling (Insight 2007)*, vol. 26 (2008), pp. 430–434
118. M. Radosavljevic, T. Ashley, A. Andreev, S.D. Coomber, G. Dewey, M.T. Emeny, M. Fearn, D.G. Hayes, K.P. Hilton, M.K. Hudait, R. Jefferies, T. Martin, R. Pillarisetty, W. Rachmady, T. Rakshit, S.J. Smith, M.J. Uren, D.J. Wallis, P.J. Wilding, R. Chau, High-performance 40nm gate length InSb p-channel compressively strained quantum well field effect transistors for low-power (VCC = 0.5 V) logic applications, in *International Electron Devices Meeting* (2008), pp. 1–4
119. L.-A. Ragnarsson, Z. Li, J. Tseng, T. Schram, E. Rohr, M.J. Cho, T. Kauerauf, T. Conard, Y. Okuno, B. Parvais, P. Absil, S. Biesemans, T.Y. Hoffmann, Ultra low-EOT (5 A) gate-first and gate-last high performance CMOS achieved by gate-electrode optimization, in *International Electron Devices Meeting* (2009), pp. 1–4
120. I. Riihimaki, A. Virtanen, S. Rinta-Anttila, P. Pusa, J. Raisanen, Vacancy-impurity complexes and diffusion of Ga and Sn in intrinsic and p-doped germanium. Appl. Phys. Lett. **91**, 091922 (2007)
121. K. Romanjek, L. Hutin, C. Le Royer, A. Pouydebasque, M.-A. Jaud, C. Tabone, E. Augendre, L. Sanchez, J.-M. Hartmann, H. Grampeix, V. Mazzocchi, S. Soliveres, R. Truche, L. Clavelier, P. Scheiblin, X. Garros, G. Reimbold, M. Vinet, F. Boulanger, S. Deleonibus, High performance 70nm gate length Germanium-On-Insulator pMOSFET with high-k/metal gate. Solid-State Electron. **53**(7), 723–729 (2009). Also in 38th European Solid-State Device Research Conference
122. E.J. Ryder, Mobility of holes and electrons in high electric fields. Phys. Rev. **90**(5), 766–769 (1953)
123. A. Satta, E. Simoen, T. Clarysse, T. Janssens, A. Benedetti, B. De Jaeger, M. Meuris, W. Vandervorst, Diffusion, activation, and recrystallization of boron implanted in preamorphized and crystalline germanium. Appl. Phys. Lett. **87**, 172109 (2005)
124. A. Satta, T. Janssens, T. Clarysse, E. Simoen, M. Meuris, A. Benedetti, I. Hoflijk, B. De Jaeger, C. Demeurisse, W. Vandervorst, P implantation doping of ge: diffusion, activation, and recrystallization. J. Vac. Sci. Technol., B **24**, 494–498 (2006)
125. A. Satta, E. Simoen, T. Janssens, T. Clarysse, B. De Jaeger, A. Benedetti, I. Hoflijk, B. Brijs, M. Meuris, W. Vandervorst, Shallow junction Ion implantation in Ge and associated defect control. J. Electrochem. Soc. **153**(3), G229–G233 (2006)
126. A. Schenk, Rigorous theory and simplified model of the band-to-band tunneling in silicon. Solid-State Electron. **36**(1), 19–34 (1993)
127. Sentaurus sdevice, ver. D-2010.03. Available from Synopsys inc. (2010)
128. Sentaurus sprocess, ver. D-2010.03. Available from Synopsys inc. (2010)
129. S. Severi, E. Augendre, S. Thirupapuliyur, K. Ahmed, S. Felch, V. Parihar, F. Nouri, T. Hoffman, T. Noda, B. O'Sullivan, J. Ramos, E. San Andres, L. Pantisano, A. De Keersgieter, R. Schreutelkamp, D. Jennings, S. Mahapatra, V. Moroz, K. De Meyer, P. Absil, M. Jurczak, S. Biesemans, Optimization of Sub-Melt Laser Anneal: Performance and Reliability, in *International Electron Devices Meeting* (2006), pp. 1–4
130. K. Shinohara, Y. Yamashita, A. Endoh, K. Hikosaka, T. Matsui, T. Mimura, S. Hiyamizu, Extremely high-speed lattice-matched InGaAs/InAlAs high electron mobility transistors with

472 GHz cutoff frequency. Jpn. J. Appl. Phys. **41**(4B), L437–L439 (2002)
131. E. Simoen, A. Brugere, A. Satta, A. Firrincieli, B. Van Daele, B. Brijs, O. Richard, J. Geypen, M. Meuris, W. Vandervorst, Impact of the chemical concentration on the solid-phase epitaxial regrowth of phosphorus implanted preamorphized germanium. J. Appl. Phys. **105**(9), 093538 (2009)
132. W. Skorupa, T. Gebel, R.A. Yankov, S. Paul, W. Lerch, D.F. Downey, E.A. Arevalo, Advanced thermal processing of ultrashallow implanted junctions using flash lamp annealing. J. Electrochem. Soc. **152**, G436 (2005)
133. C.S. Smith, Piezoresistance effect in germanium and silicon. Phys. Rev. **94**, 42–49 (1954)
134. N. Stolwijk, H. Bracht, *Landolt-Börnstein Database*, vol. III/33A (Springer, Berlin, 1998)
135. K. Suzuki, R. Sudo, Analytical expression for ion-implanted impurity concentration profiles. Solid-State Electron. **44**(12), 2253–2257 (2000)
136. K. Suzuki, Y. Tada, Y. Kataoka, Robust boron Ion implantation profile database with an energy range of 0.5 to 2000 keV based on accurate SIMS data and calibrated Monte Carlo simulation tracing to virtual negative plane, in *Proceedings of the 18th Ion Implantation Technology* (2006), pp. 2–24
137. K. Suzuki, K. Ikeda, Y. Yamashita, M. Harada, N. Taoka, O. Kiso, T. Yamamoto, N. Sugiyama, S.-I. Takagi, Ion-implanted impurity profiles in Ge substrates and amorphous layer thickness formed by Ion implantation. IEEE Trans. Electron Devices **56**(4), 627–633 (2009)
138. S.M. Sze, *Physics of Semiconductor Devices* (Wiley, Hoboken, 1981)
139. S. Takagi, A. Toriumi, M. Iwase, H. Tango, On the universality of inversion layer mobility in Si MOSFET's: Part I. Effects of substrate impurity concentration. IEEE Trans. Electron Devices **41**(12), 2357–2362 (1994)
140. T. Tanaka, Novel parameter extraction method for low field drain current of nano-scaled MOSFETs, in *International Conference on Microelectronic Test Structures* (2007), pp. 265–267
141. N. Taoka, K. Ikeda, Y. Yamashita, N. Sugiyama, S. Takagi, Effects of ambient conditions in thermal treatment for Ge(0 0 1) surfaces on Ge–MIS interface properties. Semicond. Sci. Technol. **22**(1), S114 (2007)
142. N. Taoka, M. Harada, Y. Yamashita, T. Yamamoto, N. Sugiyama, S.-I. Takagi, Effects of Si passivation on Ge metal-insulator-semiconductor interface properties and inversion-layer hole mobility. Appl. Phys. Lett. **92**(11), 113511 (2008)
143. A. Thomas, Multi-view object class recognition and metadata transfer. Ph.D. Dissertation, KU Leuven (2009)
144. M.O. Thompson, G.J. Galvin, J.W. Mayer, P.S. Peercy, J.M. Poate, D.C. Jacobson, A.G. Cullis, N.G. Chew, Melting temperature and explosive crystallization of amorphous silicon during pulsed laser irradiation. Phys. Rev. Lett. **52**(26), 2360–2363 (1984)
145. S.E. Thompson, R.S. Chau, T. Ghani, K. Mistry, S. Tyagi, M.T. Bohr, In search of "Forever", continued transistor scaling one new material at a time. IEEE Trans. Semicond. Manuf. **18**(1), 26–36 (2005)
146. S. Tian, Predictive Monte Carlo ion implantation simulator from sub-keV to above 10 MeV. J. Appl. Phys. **93**(10), 5893–5904 (2003)
147. L. Trojman, L. Pantisano, M. Dehan, I. Ferain, S. Severi, H.E. Maes, G. Groeseneken, Velocity and mobility investigation in 1-nm-EOT HfSiON on Si (110) and (100)—Does the dielectric quality matter? IEEE Trans. Electron Devices **56**(12), 3009–3017 (2009)
148. F.A. Trumbore, E.M. Porbansky, A.A. Tartaglia, Solid solubilities of aluminum and gallium in germanium. J. Phys. Chem. Solids **11**, 239–245 (1959)
149. Y. Tsividis, *Operation and Modeling of the MOS Transistor* (Oxford University Press, Oxford, 1999)
150. P. Tsouroutas, D. Tsoukalas, I. Zergioti, N. Cherkashin, A. Claverie, Diffusion and activation of phosphorus in germanium. Mater. Sci. Semicond. Process. **11**(5–6), 372–377 (2008)
151. M.S. Tyagi, R. Van Overstraeten, Minority carrier recombination in heavily-doped silicon. Solid-State Electron. **26**, 577–597 (1983)

152. S. Uppal, A.F.W. Willoughby, J.M. Bonar, A.G.R. Evans, N.E.B. Cowern, R. Morris, M.G. Dowsett, Diffusion of ion-implanted boron in germanium. J. Appl. Phys. **90**(8), 4293–4295 (2001)
153. P. Van Der Voorn, M. Agostinelli, S. Choi, G. Curello, H. Deshpande, M.A. El-Tanani, W. Hafez, U. Jalan, L. Janbay, M. Kang, K. Koh, K. Komeyli, H. Lakdawala, J. Lin, N. Lindert, S. Mudanai, J. Park, K. Phoa, A. Rahman, J. Rizk, L. Rockford, G. Sacks, K. Soumyanath, H. Tashiro, S. Taylor, C. Tsai, H. Xu, J. Xu, L. Yang, I. Young, J. Yeh, J. Yip, P. Bai, C. Jan, A 32nm low power RF CMOS SOC technology featuring high-k/metal gate, in *Symposium on VLSI Technology* (2010), pp. 137–138
154. P. Verheyen, G. Eneman, R. Rooyackers, R. Loo, L. Eeckhout, D. Rondas, F. Leys, J. Snow, D. Shamiryan, M. Demand, Th.Y. Hoffmann, M. Goodwin, H. Fujimoto, C. Ravit, B.-C. Lee, M. Caymax, K. De Meyer, P. Absil, M. Jurczak, S. Biesemans, Demonstration of recessed $Si_{1-x}Ge_x$ S/D and inserted metal gate on HfO_2 for high performance pFETs, in *International Electron Devices Meeting* (2005), pp. 886–889
155. M. Virgilio, G. Grosso, Type-I alignment and direct fundamental gap in SiGe based heterostructures. J. Phys. Condens. Matter **18**(3), 1021 (2006)
156. L. Witters, S. Takeoka, S. Yamaguchi, A. Hikavyy, D. Shamiryan, M. Cho, T. Chiarella, L.-A. Ragnarsson, R. Loo, C. Kerner, Y. Crabbe, J. Franco, J. Tseng, W.E. Wang, E. Rohr, T. Schram, O. Richard, H. Bender, S. Biesemans, P. Absil, T. Hoffmann, 8 A Tinv gate-first dual channel technology achieving low-Vt high performance CMOS, in *Symposium on VLSI Technology* (2010), pp. 181–182
157. C. Wündisch, M. Posselt, B. Schmidt, V. Heera, T. Schumann, A. Mücklich, R. Grötzschel, W. Skorupa, T. Clarysse, E. Simoen, H. Hortenbach, Millisecond flash lamp annealing of shallow implanted layers in Ge. Appl. Phys. Lett. **95**(25), 252107 (2009)
158. G. Xia, O.O. Olubuyide, J.L. Hoyt, M. Canonico, Strain dependence of Si–Ge interdiffusion in epitaxial $Si/Si_{1-y}Ge_y/Si$ heterostructures on relaxed $Si_{1-x}Ge_x$ substrates. Appl. Phys. Lett. **88**(1), 013507 (2006)
159. S. Yamaguchi, L. Witters, J. Mitard, G. Eneman, G. Hellings, M. Fukuda, A. Hikavyy, R. Loo, A. Veloso, Y. Crabbe, E. Rohr, P. Favia, H. Bender, S. Takeoka, G. Vellianitis, W. Wang, L. Ragnarsson, K. De Meyer, A. Steegen, N. Horiguchi, High performance Si.45Ge.55 implant free quantum well FET featuring low temperature process, eSiGe stressor and transversal strain relaxation, in *IEEE International Electron Devices Meeting—IEDM* (2011), pp. 829–832
160. T. Yamamoto, Y. Yamashita, M. Harada, N. Taoka, K. Ikeda, K. Suzuki, O. Kiso, N. Sugiyama, S.-I. Takagi, High performance 60nm gate length germanium p-MOSFETs with Ni germanide metal source/drain, in *International Electron Devices Meeting* (2007), pp. 1041–1043
161. J.F. Ziegler, J.P. Biersack, U. Litmark, *SRIM the Stopping Ranges of Ions in Matter* (SRIM Co., Chester, 2008)

CPSIA information can be obtained at www.ICGtesting.com
Printed in the USA
LVOW11*1228190813

348447LV00001BA/15/P